大熊猫图志

GIANT PANDA PICTORIAL ANNALS

四川省地方志工作办公室
四川省林业和草原局 编著

Edited by The Office of Sichuan Province Local Records
Sichuan Province Forestry and Grassland Administration

方志出版社
Publishing House of Local Records

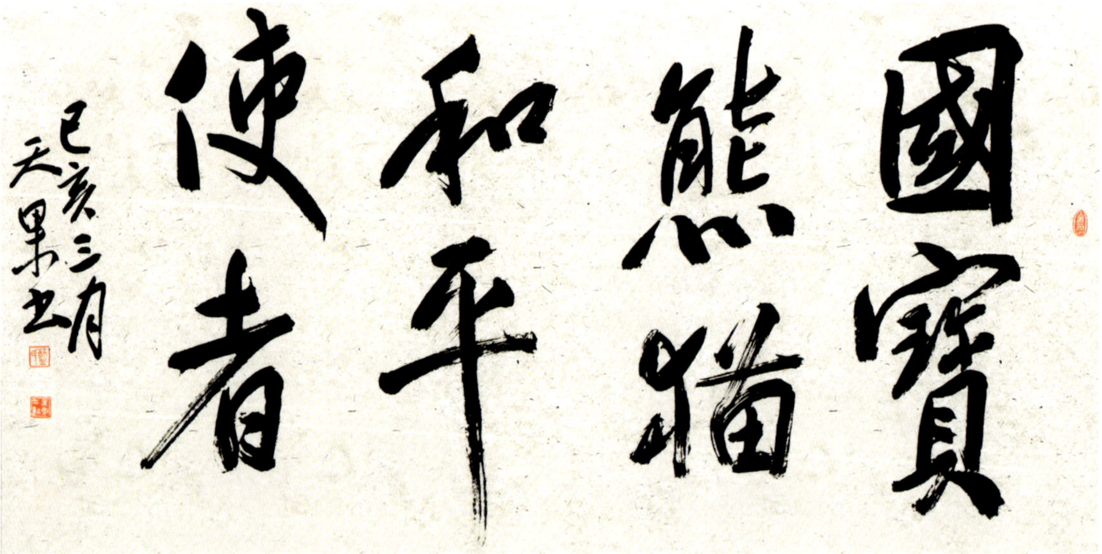

书法《国宝熊猫 和平使者》，138cm×69cm（2019 年） 蓝天果 作品

Calligraphy: *The Giant Panda – peace envoy*, 138cm×69cm (2019). By Lan Tianguo

谨以此书，纪念大熊猫科学发现150周年

In honor of the 150th anniversary of the scientific discovery of giant pandas

1869—2019

《大熊猫图志》编纂机构

一、《大熊猫图志》编纂委员会

特约顾问 胡锦矗 魏辅文

顾　　问 张和民 张志和 冉江洪 张泽钧 魏光飚 薛　康 马小彬 孙　前 周孟棋 谭　楷

主　　任 陈建春 李天满 高　山

副 主 任 赵　行 包建华 邓　瑜 陶利辉 陈宗迁 周书生

委　　员 熊　倩 高伟明 牛　森 何晓波 易红雨 吴生才 杨天明 王鸿加 张黎明 林　红 徐　宏 王晓春

二、《大熊猫图志》主编、副主编

主　　编 陈建春

副 主 编 赵　行 包建华 邓　瑜 陶利辉

三、《大熊猫图志》项目组

组　　长 张兆法

副 组 长 朱艳林 古晓东

成　　员 赵卫红 马　睿 许童春 张小军 马　玥 杨博宇 宁紫含 王小霞 孙晓雪 汪智昊 李开云 陈　兰 严青青 江林蔓 王翔宗 康　成 张　奕 李柯薪

四、《大熊猫图志》审稿小组

组　　长 张兆法 吴生才

成　　员 朱艳林 马国栋 王远森 王鸿加 杨旭煜 古晓东 林荣岗 张　倩 马　睿 许童春

五、《大熊猫图志》翻译组

组　　长 曹顺发 张小军 马　玥

副 组 长 胥　瑾 熊玥伽 王洁蕊 曹知文（美国）

成　　员 何　姗 阮先玉 李立平 王斐文 杨小武 刘　佯 郑　雪 胡新云

Compiling Team of *GIANT PANDA PICTORIAL ANNALS*

1. Editorial Board

Special Consultants	Hu Jinchu	Wei Fuwen		
Consultants	Zhang Hemin	Zhang Zhihe	Ran Jianghong	Zhang Zejun
	Wei Guangbiao	Xue Kang	Ma Xiaobin	Sun Qian
	Zhou Mengqi	Tan Kai		
Directors	Chen Jianchun	Li Tianman	Gao Shan	
Deputy Directors	Zhao Xing	Bao Jianhua	Deng Yu	Tao Lihui
	Chen Zongqian	Zhou Shusheng		
Board Members	Xiong Qian	Gao Weiming	Niu Miao	He Xiaobo
	Yi Hongyu	Wu Shengcai	Yang Tianming	Wang Hongjia
	Zhang Liming	Lin Hong	Xu Hong	Wang Xiaochun

2. Editor-in-chief, Deputy Editors-in-chief

Editor-in-chief	Chen Jianchun			
Deputy Editors-in-chief	Zhao Xing	Bao Jianhua	Deng Yu	Tao Lihui

3. Project Team

Leader	Zhang Zhaofa			
Deputy Leaders	Zhu Yanlin	Gu Xiaodong		
Members	Zhao Weihong	Ma Rui	Xu Tongchun	Zhang Xiaojun
	Ma Yue	Yang Boyu	Ning Zihan	Wang Xiaoxia
	Sun Xiaoxue	Wang Zhihao	Li Kaiyun	Chen Lan
	Yan Qingqing	Jiang Linman	Wang Xiangzong	Kang Cheng
	Zhang Yi	Li Kexin		

4. Review Panel

Leaders	Zhang Zhaofa	Wu Shengcai		
Members	Zhu Yanlin	Ma Guodong	Wang Yuansen	Wang Hongjia
	Yang Xuyu	Gu Xiaodong	Lin Ronggang	Zhang Qian
	Ma Rui	Xu Tongchun		

5. Translation Team

Leaders	Cao Shunfa	Zhang Xiaojun	Ma Yue	
Deputy Leaders	Xu Jin	Xiong Yuejia	Wang Jierui	Early Travis Patrick (USA)
Members	He Shan	Ruan Xianyu	Li Liping	Wang Feiwen
	Yang Xiaowu	Liu Yang	Zheng Xue	Hu Xinyun

凡　例

一、本图志编纂工作以马克思列宁主义、毛泽东思想、邓小平理论、“三个代表”重要思想、科学发展观、习近平新时代中国特色社会主义思想为指导，坚持辩证唯物主义和历史唯物主义的立场、观点和方法，全面客观地记述大熊猫科学发现、研究和保护的历史与现状。

二、本图志采用篇、章结构，分为六篇，每篇设有卷首语和若干章节，每一章下有正文和知识点。

三、本图志以图片和照片为主要记述内容，上限起于4000年前新石器时代末期，即考古发现的大熊猫殉葬品墓葬，下限截至2018年，以1950年到2018年为主要记述时段。但为了保持事件的完整性或反映重大事件，个别事项适当下延。图片和照片文字说明包括图所反映的主要内容、拍摄者、时间、来源等信息。记述的行政区域以四川省为主，同时兼顾陕西省和甘肃省。

四、本图志中的地名、机构等均用当时的名称记载，若名称有过变化，通过注释加以说明。机构或文件名等首次出现时用全称，并括注简称，此后一般用简称。省级人民政府、省级林业管理机构与国家部、委、局联合发文及组织有关活动时，机构名称用全称；在外事活动中机构名称用全称。科学技术术语以国家有关机关审定的为准，未经审定的遵从习惯。有关法律名称，首次出现时用全称，之后用简称，如《中华人民共和国森林法》简称《森林法》。图志中的地图，部分由相关人员绘制，部分引自其他相关著作，均符合测绘法有关规定。

五、本图志资料主要来源于现存档案和公开出版的志书、年鉴，大熊猫研究专家胡锦矗、魏辅文等的最新研究成果，以及相关单位报送及编写人员征集的材料。所用资料经过反复核对和考证，其中涉及的相关数据，以历史文献、政府公布的资料、职能部门核定的数字为准，均符合有关保密规定；对于志稿中的人物、事件，依当时习惯提法及实际情况记载。

To Our Readers

Giant Panda Pictorial Annals is compiled under the guidance of Marxism-Leninism, Mao Zedong Thought, Deng Xiaoping Theory, Important Thought of Three Represents, Scientific Outlook on Development, and Xi Jinping Thought on Socialism with Chinese Characteristics for a New Era. In line with the standpoint, views and methods of dialectical and historical materialism, it offers a comprehensive and objective account of the scientific discovery, research and conservation of the giant panda from past till now.

This book contains six major parts, each having its thesis statement and chapters, and each chapter having its text and key points.

As its title suggests, this book features illustrations and pictures, and tells stories about the giant panda from the end of the Neolithic Age 4000 years ago till the year 2018 (including some important events that have occurred in the first two quarters of 2019), focusing on what has happened in Sichuan—though Shaanxi and Gansu are mentioned from time to time—from 1950 to 2018. The reason that the stories begin with the end of the Neolithic Age is that archaeological discoveries indicate that sacrificial animals in some mausoleums in those days were shaped like the giant panda. The captions of the illustrations and pictures carry such information as their photographers, dates photographed, and or sources.

Places and institutions are recorded in their names that were in use when connected with events; when a name is no longer in use, a footnote is given. Institutions or documents first appear in full names along with abbreviations or short forms in parentheses, and abbreviations or short forms are used thereafter as a rule. Full names are used when provincial governments, provincial forestry administrations and state ministries, commissions and bureaus jointly issue documents and carry out related activities. Scientific and technical terms are either authorized ones by the relevant state organs or conventional ones. Legal terms also first appear in full names and then in abbreviations. For example, the *Forest Law of the People's Republic of China* is referred to as the *Forest Law*, for short. Some maps in this book have been drawn by experts of this compiling team; the others are quoted from relevant works. These maps are all in accordance with the Surveying and Mapping Law of People's Republic of China.

Data of this book are mainly from published archives, annals, yearbooks, latest researches of giant panda experts such as Hu Jinchu and Wei Fuwen, as well as materials provided by relevant organizations or collected by the writers. They have been checked and verified repeatedly. The statistics, used in line with the relevant provisions on confidentiality, are subject to historical documents and materials published by the government concerned. People and events are recorded according to the usage when involved.

目　录

封面摄影 野外的大熊猫（2008 年） 周孟棋 摄
封面题字 大熊猫图志（2019 年） 杨频 作品
目录底图 绢本《墨韵竹香》，70cm×138cm（2015 年） 邓维彬 作品

CONTENTS

Front cover: A Giant Panda in the Wild (2008). Photo by Zhou Mengqi

Chinese calligraphy on cover (2019). By Yang Pin

Background on the contents page: *Silk scroll: Bamboo Rhyme*, 70cm×138cm (2015). By Deng Weibin

母爱（2017 年） 周孟棋 摄

Maternal Love (2017). Photo by Zhou Mengqi

序

庆祝大熊猫科学发现 150 周年

这本图志是我们在大熊猫科学发现 150 周年之际，为全世界准备的一份礼物，也是对所有曾经或正致力于大熊猫保护的人们送上的一份祝福。正是因为大家的努力，人类和大熊猫才彼此安好！

在历史长河里，150 年不过弹指一挥间。但是，正是这百余年时间，让大熊猫从深山峡谷走向全世界，成为全球超级明星。透过大熊猫这个“活化石”，我们有幸窥探古老的地球奥秘，并更加重视生态文明。

1985 年，世界野生动植物基金会（WWF）宣布大熊猫为世界十种濒危物种之一。但是 2016 年，世界自然保护联盟（IUCN）宣布将大熊猫受威胁程度从“濒危”降为“易危”。中国四次大熊猫调查结果显示，野生大熊猫数量已从减少变成稳中有升。这些消息让人们振奋。这既是一份天赐的美好，也是对全人类为大熊猫繁衍生息而努力的回馈。

大熊猫以黑白两色示人，生活极简，憨态可掬，它宽厚、祥和的形象和持续不断的进化，象征着中华民族的文化与精神。在这一古老而充满希望的物种身上，我们看到大自然的勃勃生机。

针对大熊猫的研究和保护依然在不断深入，来自全世界的工作者正把这项事业提升到崭新的高度。作为大熊猫的故乡，中国重任在肩。

值此大熊猫科学发现 150 周年之际，在四川省人民政府新闻办公室指导下，四川省地方志工作办公室、四川省林业和草原局联合编著出版全球第一部全方位记录大熊猫的图志。图志中的每一张图片、每一段文字，都饱含了人类对大熊猫的大爱。希望读者在阅读这本图志时，能够充满爱意并触摸到生生不息的希望。

大熊猫来自中国，也属于全世界。我们欢迎所有关心大熊猫的朋友，持续为保护大熊猫而努力！

未来我们继续同行！

是为序！

章玉钧

2019 年 6 月 28 日

Foreword

In honor of the 150th anniversary of the scientific discovery of giant pandas

Giant Panda Pictorial Annals, as our gift to the world to mark the 150th anniversary of the scientific discovery of the giant panda, conveys our best wishes to all those who have been or are still committed to the conservation endeavor of this species. Thanks to everyone's efforts, man and the giant panda co-exist in peace and harmony!

150 years are only a blink of an eye in the long history of the earth. However, this fleeting span of time has witnessed how the giant panda have gone from the bamboo groves deep in the mountains in western China to metropolises in the world. They have now become global superstars, with ever-increasing popularity around the world. By studying this "living fossil", we can unravel some mysteries of our ancient earth and realize the importance of building an ecological civilization.

In 1985, World Wide Fund for Nature (WWF) announced the giant panda one of the top ten endangered species on our planet. Then in 2016, the International Union for Conservation of Nature (IUCN) announced that the status quo of the giant panda has been improved from "endangered" to "vulnerable." According to the four national censuses of the giant panda in China (1974-77, 1985-88, 1999-2003, 2011-14), their population in the wild has been on the rise. This growth is a heartening message and a blessing to lovers of this species, and a reward to our conservation efforts.

The black-and-white giant panda live but a simple life. Their leisurely and peaceful look, constant evolution and adaptation to the changing environment symbolize the Chinese civilization and spirit. From this species that has come all the way from remote antiquity, we have seen the extraordinary vitality of nature.

Significant achievements have been made in scientific research and conservation of the giant panda. Indeed, biologists and animal conservationists around the globe have vigorously been pushing forward the conservation endeavor. As home to the giant panda, China is obligated to take care of them.

To mark the 150th anniversary of the giant panda's scientific discovery, Sichuan Province Local Records Office and Sichuan Province Forestry and Grassland Administration have jointly compiled and published *Giant Panda Pictorial Annals*—the most comprehensive of all those devoted to the depiction of the giant panda—under the guidance of Information Office of Sichuan Provincial People's Government. The pictures, maps, and narratives in it carry our empathy and sympathy for this species. We expect that you, dear readers, feel the same and have strong faith in our efforts.

The giant panda is native to China, and belongs to the rest of the world, too. Their conservation is your obligation as well as ours. Let's join hands to support this vulnerable species for their continuation on this planet.

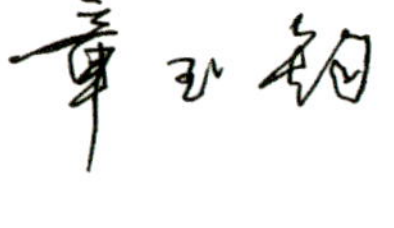

June 28, 2019

大熊猫精神

伏羲演八卦以观于天，庄子以太极喻宇宙无极。《道德经》说：“道生一，一生二，二生三，三生万物。万物负阴而抱阳，冲气以为和。”大熊猫黑白两色是阴阳的象征，不仅契合中国人最古朴的审美习惯，也表现出中国人最古老的哲学智慧。

大熊猫洒脱率真，不疾不徐，迈着内八字在竹林里缓缓踱步。这些“竹林隐士”有着自己的“气节”，它们挑选最可口的食材，然后优雅地品尝。大熊猫对水源质量也很“讲究”，跋涉几座山也要寻到清泉溪流才肯畅饮。这样的性格，更像是超凡脱俗的“高人逸士”，有所为而有所不为。

大熊猫性格坚韧，应对了一次又一次冷暖交替的恶劣环境，成为抗击冰川的战士。无论酷暑严寒，它们都“遗世独立”。恶劣的环境没有改变它们善良、率真的秉性，在困境中百折不挠的精神是中国人自古而今的傲然风骨。

大熊猫从不“惹是生非”，在野外遇到不友善的家伙也多采取躲避的态度，更不主动攻击其他动物。大熊猫的邻居很多，它们和谐相处，共享大自然赋予的资源。

外圆内方、刚柔并济、敦厚友善、爱好和平，这正是中国的外交形象，也是中国人的精神内核。正因如此，大熊猫始终作为“友好使者”推动着中国与世界的友谊发展。

穿过岁月长河，作为适者生存的代表，大熊猫缓缓向我们走来。它顺应自然，与自然和谐共生。这种顺势而为、追求和谐的精神，与中国传统的民族精神不谋而合，同时还契合改善生态环境、维护生态平衡这一全人类的共同主题。

作为中国独有的物种之一，大熊猫身上有许许多多中华美德：憨厚、可爱、友善、和平；强健、沉稳、包容、应变。“大熊猫精神”连接历史，纵横现代，延至国际，展现的正是中国精神。这种精神无时空、无国界、无民族，为世界所接受，为人类所尊崇。

四川省地方志工作办公室
四川省林业和草原局
2019年6月28日

Preface

The Spirit of the Giant Panda

Fu Hsi, one of the earliest sage sovereigns in China, created Eight-trigrams (*ba-gua*) to probe into the mysteries of heaven and earth; Chuang Tzu, a famous philosopher in China, founded a theory in which the Great Ultimate (*tai-ji*) is applied to interpret the Infinite of the Universe (*wu-ji*). There is a saying in *Tao Te Ching* written by Lao Tzu, "One is the child of the divine law. After one come two, after two come three, after three come all things. Everything has its bright and dark side, co-existent in harmony." The black-and-white color of the giant panda is a symbol of the balance between Yin and Yang. It echoes the traditional Chinese aesthetic values and embodies the philosophical wisdom of the Chinese ancients.

The giant panda lives an easy and leisurely life, strolling with pigeon toes in the bamboo forests at their own pace, and picking and choosing the most delicious food, nibbling and crunching unhurriedly. Creatures of this species are also particular about their drinking water, for, to quench thirst, they will trudge one mountain after another in search of a clear spring or stream. With such a trait, they resemble human hermits who do things after their own hearts.

They have great adaptability, which enables them to survive harsh environmental changes and defy the challenges of glaciers. However severe the weather, they just brave it on their own feet. The indomitable spirit in face of any predicament is indeed what has been supporting the Chinese people from time immemorial to the present.

No giant panda is a "troublemaker," never initiating its attack on any other species. When threatened, they tend to retreat, and avoid fighting; they share their habitat peacefully with all other species.

In a word, the giant panda is gentle and firm at the same time, friendly and peaceful. These traits, as the cultural core of the Chinese people, are all embodied in their diplomacy. For this very reason, the giant panda plays the role of a goodwill ambassador to many other countries in the world; they are promoters of friendly exchanges between China and the rest world.

Through the long history of evolution, the giant panda has proved to be a role model in terms of the survival of the fittest, for they have adapted themselves to the changes of nature and managed to produce and reproduce until today with the help of human efforts. The spirit of making the best of and pursuing harmony with the environment is also the spirit of the Chinese people. The survival experience of the giant panda echoes the common theme of mankind-improving ecological environment and maintaining ecological balance.

As a rare species that only lives deep in the

mountains in China's Sichuan, Shaanxi and Gansu, the giant panda represents numerous virtues of the Chinese people: friendliness, peace-loving, valiance, adaptability, resilience, and composure in the face of danger. Indeed, these are time-honored virtues in China and throughout the world.

The Office of Sichuan Province Local Records
Sichuan Province Forestry and Grassland Administration
June 28, 2019

绢本《暖秋》，135cm×213cm（2017 年） 王申勇 作品
Ink and color on silk: *On a Warm Autumn Day*, 135cm×213cm (2017). By Wang Shenyong

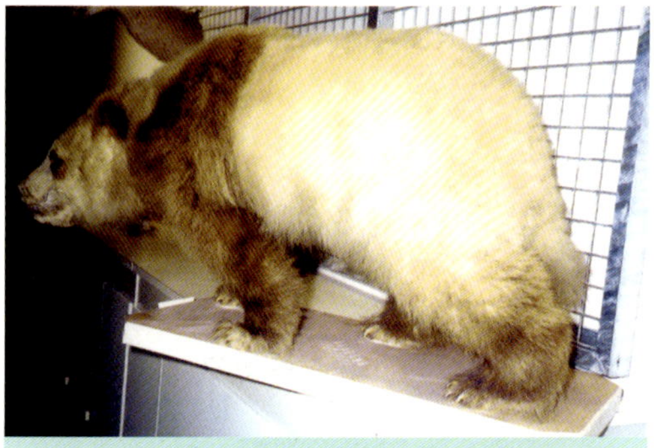

清同治八年（1869 年）

4 月 1 日，法国博物学家阿尔芒·戴维神父在中国四川省穆坪（今宝兴县）邓池沟科学发现大熊猫，并把它初步定名为“黑白熊”。

清同治十一年（1872 年）

巴黎自然历史博物馆主任米勒·爱德华将“黑白熊”重新定名为 *Ailuropus melanoleucus*（拉丁名，意思是“黑白相间的熊猫”）。

清光绪二十七年（1901 年）

英国大英博物馆主任兰克斯特在研究了保存于该馆的两具大熊猫标本后发表论文，并将现代大熊猫英文名定为 Giant Panda。

民国 21 年（1932 年）

中国动物学界推测大熊猫主要分布在中国陕西省秦岭山脉以南和西藏自治区以东到云南省北部地区。

民国 25 年（1936 年）

11 月，美国女服装设计师露丝·哈根纳斯在中国四川省威州（今汶川县）捕获到大熊猫“苏琳”。1937 年 2 月，“苏琳”在美国芝加哥动物园展出。这是活体大熊猫第一次被带出中国。

民国 28 年（1939 年）

8 月 11 日，一只大熊猫从中国四川省成都华西协合大学[1]被运送到重庆北碚平民公园圈养并展出。这是中国动物园首次圈养大熊猫。

[1] 清宣统二年（1910 年），美国、英国、加拿大的 5 个教会组织在成都华西坝建立华西协合大学。该校 1951 年改名为华西大学，1953 年改名为四川医学院，1985 年改名为华西医科大学，2000 年并入四川大学。

民国 30 年（1941 年）

为感谢“美国援华联合会”对中国抗日战争的援助，宋美龄将大熊猫“潘弟”“潘达”赠送给美国人民。

1953 年

成都动物园首次饲养、展出大熊猫幼崽。

1955 年

5 月，四川省捕捉到大熊猫“平平”。后被作为国礼送到苏联莫斯科动物园，成为中华人民共和国第一只外送大熊猫。

1959 年

4 月 1 日，大熊猫的形象第一次出现在中华人民共和国邮电部发行的中央自然博物馆特种邮票上。该邮票 1 套 2 枚，由邮票设计家吴建坤设计，面值分别为 4 分和 8 分。

1961 年

11 月，世界野生动植物基金会（WWF）[2]成立，并以大熊猫图像作为该组织的会旗和会徽图案。

1963 年

4 月 2 日，中国四川省建立汶川县卧龙、平武县王朗、南坪县（今九寨沟县）白河、天全县喇叭河等自然保护区。这是中国首批建立的以大熊猫及其栖息地为主要保护对象的自然保护区。

[2] 1986 年，世界野生动植物基金会“World Wildlife Fund”，改名为世界自然基金会“World Wide Fund For Nature”。本图志中该组织英文名统一使用“World Wide Fund For Nature”。

北京动物园的大熊猫“森森”和“莉莉”进行自然交配，成功产下一崽。这是世界上第一例人工繁育的大熊猫。

1964 年

北京师范大学郑光美、徐平宇发表《秦岭南麓发现的大猫熊》一文，首次公开报道在中国陕西省南部发现大熊猫。其后，中国陕西省佛坪县大熊猫开始为世人所瞩目。

1968 年

中国科学院动物研究所在平武县王朗自然保护区开展大熊猫野外数量调查。这是首次针对大熊猫的调查活动。

1972 年

中国先后向美国和日本分别赠送两只大熊猫。

1977 年

10 月，全国第一次大熊猫调查结束。据估算，中国约有野生大熊猫 2459 只。

1978 年

3 月，南充师范学院胡锦矗和有关研究人员在卧龙自然保护区建立世界上首个大熊猫野外生态观察站“五一棚”。80 年代中期以后，唐家河白熊坪、马边大风顶暴风坪等野外观察站先后成立，为探索大熊猫生活的科学奥秘奠定了基础。

9 月 8 日，北京动物园大熊猫“涓涓”一胎产下两崽。这是世界上第一次采用人工授精方法繁殖大熊猫成功的记录。

12 月 15 日，国务院批转国家林业总局《关于加强大熊猫保护、驯养工作的报告》，提出在四川省的大风顶、唐家河、九寨沟和陕西省岳坝再划建 4 个自然保护区。同时，在四川省卧龙、甘肃省白水江和陕西省岳坝建立 3 个大熊猫饲养繁殖中心。

1979 年

12 月 28 日，林业部批准建立佛坪自然保护区，1986 年更名为陕西佛坪自然保护区。

1980 年

5 月，四川卧龙国家级自然保护区被联合国教科文组织（UNESCO）列入世界生物圈保护区网，并与世界野生动植物基金会合作，在卧龙建立了中国保护大熊猫研究中心。1983 年研究中心建成并投入使用。

12 月 25 日，中国正式加入《濒危野生动植物种国际贸易公约》（CITES）。1981 年 4 月 8 日起，该公约对中国生效。

成都动物园首次采用雄性大熊猫“珠珠”的冻精（收集于 1979 年）为雌性大熊猫“美美”授精，产下“蓉生”。这是世界第一例利用冻精繁殖的大熊猫。

1983 年

3 月，四川省决定成立汶川县卧龙特别行政区，简称“卧龙特区”。

7 月 22 日，为哀悼出生 3 个小时后死去的大熊猫幼崽，世界野生动植物基金会在瑞士格朗总部全天下半旗。这一做法属世界首次。

1985 年

2 月，世界野生动植物基金会发表报告，把大熊猫列为世界十种濒危动物之一。

3 月 26 日，中国陕西省佛坪县发现第一只毛色棕白相间的患病大熊猫。这只大熊猫经抢救治疗恢复健康，被取名为“丹丹”。

1986 年

7 月 9 日，国务院批准唐家河自然保护区等 20 个自然保护区为第一批国家级森林和野生动物类型自然保护区。

1987 年

成都大熊猫繁育研究基地成立。

1988 年

11 月，全国第二次大熊猫调查结束。结果显示，中国有野生大熊猫 1114 只。

1989 年

1 月，中国大熊猫繁育技术委员会成立。

8 月 31 日，第一只棕白色大熊猫“丹丹”产下一崽，毛色黑白，取名“秦秦”。

1990 年

8 月，成都动物园的大熊猫“庆庆”产下一对双胞胎“娅娅”“祥祥”，并在人工辅助下成活。这是世界上第一例人工辅助哺育成活的大熊猫双胞胎。

9 月，大熊猫“盼盼”成为北京亚运会吉祥物。

1992 年

9 月，成都动物园与成都大熊猫繁育研究基地人工辅助大熊猫哺育一胎二崽成活，同时该母亲还代育异母一崽，创造了大熊猫育幼史上一母育三崽的成活奇迹。

1993 年

9 月 24—28 日，首届国际熊猫节在中国四川省成都市举办。

9 月，成都大熊猫繁育研究基地建成世界第一家大熊猫主题博物馆。

1999 年

3 月 11 日，中央人民政府向香港特别行政区赠送一对大熊猫“安安”和“佳佳”。

8 月 18 日，中国保护大熊猫研究中心大熊猫“1 号”一胎产三崽，成活两崽。这是人工饲养大熊猫以来仅此一例的三胞胎。

2000 年

11 月，中共四川省委、四川省人民政府出台《关于加快培育旅游支柱产业建设旅游经济强省的决定》（川委发〔2000〕48 号），提出以国宝大熊猫、世界自然遗产九寨沟以及古蜀文化瑰宝三星堆为旗帜，树立起四川旅游的国际品牌形象。

2003 年

4 月，世界自然基金会授予陕西省人民政府“献给地球的礼物”荣誉证书，以表彰其在秦岭大熊猫保护工作中的突出贡献。

7 月，卧龙中国保护大熊猫研究中心启动了圈养大熊猫野化培训与放归项目。

12 月，全国第三次大熊猫调查结束。结果显示，中国有野生大熊猫 1596 只。

成都大熊猫繁育研究基地旅日大熊猫“梅梅”同时自然哺育双胞胎成活，这在圈养条件下尚属全球首例。

2004 年

成都大熊猫繁育研究基地建立全球首个大熊猫基因资源库。

2005 年

6 月，受国家林业局野生动植物保护司委托，四川省野生动物资源调查保护管理站编制完成《全国大熊猫野化放归总体规划（2005—2014 年）》。

8 月 8 日，国家林业局与四川省人民政府在都江堰龙溪—虹口国家级自然保护区将救护的大熊猫“盛林 1 号”放归野外，这是中国首只正式放归野外的救护野生大熊猫。

11 月 11 日，大熊猫“晶晶”被选为 2008 年北京奥运会吉祥物。

2006 年

4 月 28 日，中国保护大熊猫研究中心人工圈养大熊猫“祥祥”放归野外。这是中国首次对人工繁育大熊猫进行野外放归试验。

7 月 12 日，在立陶宛举行的第 30 届世界遗产大会上，全票通过中国申报的“四川大熊猫栖息地”为世界自然遗产。

8 月 7 日，成都大熊猫繁育基地超轻早产儿大熊猫“五一”出生。刚出生时的体重仅 51 克，所以得名“五一”。它是世界上初生体重最轻并存活下来的大熊猫。

10 月 25 日，因在大熊猫和生物多样性保护方面做出突出贡献，世界自然基金会授予四川省人民政府、甘肃省人民政府“献给地球的礼物”证书。

2007 年

6 月 30 日，为庆祝香港回归祖国 10 周年，中央人民政府向香港特别行政区赠送一对大熊猫“盈盈”和“乐乐”。

2008 年

5 月，在“5·12”汶川特大地震中，中国保护大熊猫研究中心受损严重，幸存圈养大熊猫被转移至成都大熊猫繁育研究基地等中国各大熊猫养殖基地。

12 月 23 日，中国大陆赠送台湾一对大熊猫“团团”和“圆圆”。

2009 年

4 月 29 日，科研人员将大熊猫救护的“泸欣”移地放归至四川省石棉县栗子坪自然保护区。3 年后，“泸欣”产下幼崽。这标志着以移地放归的方式促进大熊猫小种群复壮的探索取得成功。

2010 年

12 月 18 日，中央人民政府向澳门特别行政区赠送一对大熊猫“开开”和“心心”，定居于澳门石排湾郊野公园。2016 年，它们繁育成活大熊猫双胞胎“健健”和“康康”。

2012 年

10 月 11 日，全球首只经母兽带崽野化培训的人工繁育大熊猫“淘淘”在中国四川省石棉县栗子坪自然保护区放归自然。

2013 年

11 月 6 日，大熊猫“张想”被放归到中国四川省石棉县栗子坪自然保护区。这是全球首只放归野外的圈养雌性大熊猫。

2015 年

2 月 28 日，全国第四次大熊猫调查结果公布。截至 2013 年底，中国野生大熊猫种群数量为 1864 只。

2016 年

4 月，按照中央安排，四川牵头，与陕西、甘肃启动大熊猫国家公园体制试点方案编制工作。

10 月 20 日，人工繁育大熊猫“华妍”“张梦”在中国四川省石棉县栗子坪国家级自然保护区放归自然。这是全球首次同时放归两只大熊猫。

12 月 5 日，中共中央总书记习近平主持召开中央深化改革领导小组会议，审议通过《大熊猫国家公园体制试点方案》，中国大熊猫保护事业跨入以国家公园为主体的新时代。

2017 年

3 月，圈养大熊猫“草草”与野生大熊猫实现自然交配，标志着全球范围内大熊猫首次野外引种试验取得初步成效。

5 月，受国家林业局野生动植物保护与自然保护区管理司委托，四川省野生动物资源调查保护管理站与中国科学院动物研究所等单位合作编制完成《全国大熊猫野化放归总体规划（2017—2026 年）》。

2018 年

10 月 29 日，大熊猫国家公园管理局在四川成都成立。国家林业和草原局局长张建龙和中共四川省委副书记、四川省人民政府省长尹力等共同为管理局揭牌。

11 月 8 日，中共四川省委书记彭清华在香港以“三九大”作为关键词介绍四川，“三”“九”“大”分别指三星堆、九寨沟、大熊猫，这是四川悠久的历史文化和丰富资源禀赋的集中体现。

11 月，大熊猫国家公园四川、陕西、甘肃三省管理局陆续挂牌。

2019 年

1 月 15 日，大熊猫国家公园四川省管理局成都市、绵阳市、雅安市、广元市、阿坝藏族羌族自治州、德阳市、眉山市 7 个管理分局挂牌成立。

1 月，中国第一部以大熊猫为主要记述内容的特色专志《四川省志·大熊猫志》正式公开发行。6 月，四川省地方志工作办公室、四川省林业和草原局联合编著出版《大熊猫图志》。

3 月 8 日，十三届全国人大二次会议举行开放日活动。全国人大代表、中共四川省委书记彭清华手拿熊猫冰箱贴，介绍大熊猫。

3 月 10 日，十三届全国人大二次会议四川代表团全体会议召开，讨论通过了以四川代表团名义提出的 13 件建议，其中包括：关于调整优化自然保护区和永久基本农田范围的建议，关于加大大熊猫国家公园建设政策资金支持力度的建议。

On April 1,1869, French naturalist Fr Jean Pierre Armand David discovered the giant panda in Dengchigou in Muping County (now Baoxing County) in Sichuan Province and named it "black-and-white bear."

In 1872, Milne Edwards, director of National Museum of Natural History in Paris, renamed it as *Ailuropus melanoleucus* (meaning black-and-white bear in Chinese).

In 1901, Ray Lankester, director of the British Museum, published an article after researching on the specimens of two giant pandas kept in the museum and officially named it Giant Panda.

In 1932, Chinese zoologists speculated that giant pandas were mainly distributed south of Qinling Mountains in Shaanxi Province and in the areas from the east of Tibet to the north of Yunnan Province.

In November 1936, American costume designer Ruth Harkness captured a giant panda in Weizhou (now Wenchuan County), Sichuan Province, and named it Su Lin. In February 1937, Su Lin made its debut in Chicago Zoo in the United States. It was the first time that a live giant panda had been taken out of China.

On August 11,1939, a giant panda was taken from West China Union University[1] in Chengdu, Sichuan to be kept in captivity in Beibei Civil Park in Chongqing. It was the first time that a Chinese zoological park tried keeping a giant panda in captivity.

In 1941, Song May-ling presented two giant pandas, Pandee and Pandah, to the American people as a token of China's gratitude for the support of the American Federation of Aid during the War of Resistance against Japanese Aggression.

[1]In 1910, five pastoral ministers from America, Britain and Canada established West China Union University in Huaxiba, Chengdu. In 1951, it was renamed as West University and in 1953 as Sichuan Medical College. In 1985, it was renamed as West China University of Medical Science. In 2000, it was merged into Sichuan University.

In 1953, Chengdu Zoo began to keep and exhibit giant panda cubs for the first time.

In May 1955, the giant panda Ping Ping was captured in Sichuan, and later was sent to a Moscow Zoo as a national present. It was the first giant panda sent abroad by the People's Republic of China.

On April 1, 1959, the giant panda appeared for the first time in the special stamps of Central Museum of Natural History of China issued by Ministry of Posts and Telecommunications. This set of stamps, designed by Wu Jiankun, a famous postage stamp designer, consisted of two stamps with respective face values of ¥ 0.04 yuan and ¥ 0.08 yuan.

In November, 1961, World Wide Fund for Nature (WWF)[2] was founded and adopted the image of the giant panda as its flag and emblem.

On April 2, 1963, Sichuan Province established several nature reserves including Wolong Nature Reserve in Wenchuan County, Wanglang Nature Reserve in Pingwu County, Baihe Nature Reserve in Nanping County(now Jiuzhaigou) and Labahe Nature Reserve in Tianquan County. These are the first nature reserves in China to protect the giant panda and its habitat.

In 1963, a giant panda cub was born in Beijing Zoo; it was the result of the natural mating of Sen Sen (male) and Li Li (female). This was the world's first giant panda born in captivity.

In 1964, Zheng Guangmei and Xu Pingyu from Beijing Normal University published an article entitled "Cat Bears Found at the Southern Foot of Qinling Mountains", reporting for the first time the giant panda's discovery in southern Shaanxi, China. Since then, the giant panda in Foping County, Shaanxi Province has begun to capture the world's attention.

[2]In 1986, World Wildlife Fund was changed to be World Wide Fund for Nature. This book adopts World Wide Fund for Nature or WWF.

In 1968, Zoology Institute under Chinese Academy of Sciences (CAS) initiated China's first census of the giant panda population in the wild in Wanglang Nature Reserve, Pingwu County.

In 1972, China sent two giant pandas to the USA and two others to Japan.

In October, 1977, China's First National Census of the Giant Panda was completed, which estimated that there were about 2,459 individuals in the wild.

In March, 1978, the world's first observation station of wild giant pandas, Wuyipeng, was established by Hu Jinchu (from Nanchong Teachers' College) and other researchers in Wolong National Reserve. Since the mid-1980s, such observation stations as Baixiongping Observation Station in Tangjia River Nature Reserve and Baofengping Observation Station in Mabian Dafengding Nature Reserve have been established, which constitute bases for the exploration of the survival mystery of giant pandas.

On September 8, 1978, the giant panda Juan Juan in Beijing Zoo gave birth to twin cubs. It was the first successful case of artificial insemination of the giant panda in the world.

On December 15, 1978, the State Council endorsed the "Report on Strengthening the Protection and Domestication of the Giant Panda" submitted by the State Forestry Administration, which proposed the construction of another four nature reserves in Dafengding, Tangjiahe and Jiuzhaigou in Sichuan Province and Yueba in Shaanxi Province in addition to three breeding centers for the giant panda to be built respectively in Wolong in Sichuan, Baishuijiang in Gansu, and Yueba in Shannxi.

On December 28, 1979, the Ministry of Forestry approved the establishment of Foping Nature Reserve, which was renamed in 1986 as Foping Nature Reserve in Shaanxi.

In May, 1980, Wolong National Nature Reserve in Sichuan was listed in the global network of biosphere reserves by United Nations Educational, Scientific, and Cultural Organization (UNESCO). It also cooperated with WWF to establish China Conservation and Research Center for the Giant Panda. In 1983 the center was completed and put into use.

On December 25, 1980, China officially joined the Convention on International Trade in Endangered Species (CITES), which came into effect in China on April 8,1981.

In 1980, for the first time, Chengdu Zoo used the frozen sperm of giant panda Zhu Zhu (collected in 1979) to fertilize giant panda Mei Mei, who later gave birth to Rong Sheng, which became the first in the world reproduced from frozen sperm.

In March, 1983, Sichuan Provincial Government decided to establish Wolong Special Administrative Region in Wenchuan County.

On July 22, 1983, WWF unprecedentedly lowered its flag at its headquarters in Gland, Switzerland, to mourn the death of a giant panda cub three hours after its birth.

In February, 1985, WWF issued a report and listed the giant panda as one of the top ten endangered animals in the world.

On March 26, 1985, a brown-and-white sick giant panda was found in Foping County, Shaanxi. After proper treatment, it recovered and was named Dan Dan. This was the first brown-and-white giant panda discovered in the wild.

On July 9, 1986, the State Council approved twenty nature reserves, including Tangjiahe Nature Reserve among the first batch of national forest and wild animal nature reserves.

In 1987, Chengdu Research Base of Giant Panda Breeding (Chengdu Panda Base, for short) was founded.

In November,1988, China's Second National Census of the Giant Panda was completed, which showed that there were 1,114 across China.

In January, 1989, Chinese Committee of Breeding Techniques for the Giant Panda was founded.

On August 31, 1989, Dan Dan, the first brown giant panda that had been found in the world, birthed its cub Qin Qin.

In August, 1990, giant panda Qing Qing in Chengdu Zoo gave birth to twin cubs Ya Ya and Xiang Xiang, the first that survived through human intervention.

In September, 1990, giant panda Pan Pan became the mascot of the Asian Games in Beijing.

In September, 1992, Chengdu Zoo and Chengdu Panda Base applied the artificial breeding technique to help a female panda deliver twin babies, both of whom survived. It was a miracle in the history of artificial breeding of the giant panda where the mother, while nursing its twin babies, suckled one more born by another.

On September 24 through to 28, 1993, the first International Panda Festival was held in Chengdu, China.

In September, 1993, Chengdu Panda Base established and completed the first Giant Panda Theme Museum in the world.

On March 11, 1999, the Central People's Government sent a pair of giant pandas named An An and Jia Jia to Hong Kong Special Administrative Region.

On August 18, 1999, giant panda Yi Hao, who lived in the China Panda Center, gave birth to triplets and two survived. It was the only case of three cubs at one birth in the history of artificial breeding of this species.

In November, 2000, Sichuan Provincial Committee of the Communist Party of China and the People's Government of Sichuan Province issued The Decision on Making Tourism A Pillar Industry and Building A Strong Tourist Economy (Document No. 48 issued by Sichuan Provincial Committee of the Communist Party of China in 2000). It proposed to take the giant panda (China's national treasure), Jiuzhaigou (the world natural heritage), and Sanxingdui Museum (a historical site of ancient Shu Civilization) as flags of tourism in Sichuan.

In April 2003, WWF awarded the Shaanxi Provincial People's Government the certificate of "Gift to the Earth" in recognition of its outstanding contribution to the protection of giant pandas in Qinling Mountains.

In July 2003, China Conservation and Research Center for the Giant Panda at Wolong launches the project of survival training and reintroduction of captive giant pandas.

In December, 2003, China's Third National Census of the Giant Panda was completed, which showed that there totaled 1,596 wild giant pandas in China.

In 2003, Mei Mei, which had come from Chengdu Panda Base, birthed two cubs in Japan, and managed to breastfeed them. This was the first successful case of its kind in captive conditions.

In 2004, Chengdu Panda Base established the first giant panda gene bank in the world.

In June, 2005, *The Master Plan for Giant Panda Reintroduction (2005-2014)* was formulated by Sichuan Station for Wildlife Resources Investigation and Protection Management, which had been under the supervision of both Department of Wildlife and Forest Plants Protection with the State Forestry Administration.

On August 8, 2005, State Forestry Administration and People's Government of Sichuan Province returned Shenglin 1 to the wild in Longxi-Hongkou National Nature Reserve in Dujiangyan, marking the official reintroduction of the first rescued wild giant panda in China.

On November 11, 2005, giant panda Jing Jing was selected as one of the mascots for 2008 Beijing Olympic Games.

On April 28, 2006, Xiang Xiang, kept in captivity in the China Panda Center in Wolong, was reintroduced to the wild as the first trial of releasing the artificially bred giant panda.

On July 12, 2006, the application of the Sichuan Giant Panda Habitat as a world heritage by China was passed unanimously at the 30th World Heritage Convention held in the Republic of Lithuania.

On August 7, 2006, Wu Yi, a premature cub was born in Chengdu Panda Base, weighing only 51g, then the name Wu Yi (meaning 51 in Chinese). It is the lightest cub that has survived in the the history of panda conservation.

On October 25, 2006, WWF awarded the certificates of "Gift To the Earth" to the People's Government of Sichuan and that of Gansu for their prominent contributions to the giant panda's protection and biodiversity conservation.

On June 30, 2007, the Central People's Government sent a pair of giant pandas Ying Ying and Le Le to Hong Kong Special Administrative Region in celebration of the 10th anniversary of its return to China.

On May 12, 2008, a violent earthquake that hit Wenchuan seriously damaged the China Panda Center. Those giant pandas alive were transferred to Chengdu Panda Base and other breeding bases.

On December 23, 2008, a pair of giant pandas Tuan Tuan and Yuan Yuan were sent to Taiwan by mainland China.

On April 29, 2009, researchers reintroduced giant panda Lu Xin to Liziping National Nature Reserve in Shimian County, Sichuan, which is not its native habitat. Three years later, Lu Xin birthed a cub, which means a successful trial of revitalizing a small population by returning a captive panda to a place other than her native habitat.

On December 18, 2010, giant pandas Kai Kai and Xin Xin, gifts from the Central People's Government to Macao Special Administrative Region, got settled in the Country Park of Shek Pai Wan. In 2016, they reproduced twin cubs Jian Jian and Kang Kang.

On October 11, 2012, Tao Tao, an artificially bred giant panda and the first which had received survival training in its mother company, was released to the wild in Liziping National Nature Reserve in Shimian County, Sichuan.

On November 6, 2013, giant panda Zhang Xiang was reintroduced to Liziping National Nature Reserve in Shimian County, Sichuan. It was the first captive female giant panda released to the wild.

On February 28, 2015, China's Fourth National Census of the Giant Panda revealed that by the end of 2013 there were 1,864 wild ones in China.

In April, 2016, in accordance with the arrangement of the Central Government of China, the three provinces of Sichuan, Shaanxi and Gansu launched the pilot Program of Giant Panda National Park System.

On October 20, 2016, Hua Yan and Zhang Meng, two artificially bred giant pandas, were reintroduced to the wild in Liziping National Nature Reserve, Shimian County, Sichuan. It was the first time that two giant pandas were reintroduced at the same time.

On December 5, 2016, Xi Jinping, General Secretary of the CPC Central Committee, hosted a conference for the Leading Group for Continuing the Reform, and *The Pilot Program of Giant Panda National Park System* was approved at this conference, which means the conservation of the giant panda has entered into a new era.

In March, 2017, Cao Cao, a captive giant panda, mated with a wild male naturally, marking the initial success in introducing wild giant panda gene into reproduction for the first time.

In May, 2017, *The Master Plan for Giant Panda Reintroduction (2017-2026)* was formulated by Sichuan Station for Wildlife Resources Investigation and Protection Management, along with Institute of Zoology, Chinese Academy of Sciences, and other organizations concerned. The formulation of the document had been under the supervision of both Department of Wildlife and Forest Plants Protection under the State Forestry Administration.

On October 29, 2018, Giant Panda National Park Administration was founded in Chengdu, Sichuan Province. Zhang Jianlong, director of China's Administration of Forestry and Grassland, and Yin Li, Governor of Sichuan Province unveiled the plaque for the administration.

On November 8, 2018, when Peng Qinghua, Secretary of Sichuan Provincial CPC Committee, visited Hong Kong, he introduced Sichuan with three key words: San (three in English), Jiu (nine in English) and Da (giant in English), which respectively means Sanxingdui Museum, Jiuzhaigou and Giant Panda. They embody the long history, colorful culture and rich natural resources of Sichuan Province.

In November, 2018, Giant Panda National Park Administration was established respectively in Sichuan, Shaanxi and Gansu.

On January 15, 2019, seven branches of Giant Panda National Park Administration were founded in Chengdu, Mianyang, Ya'an, Guangyuan, Aba Tibetan and Qiang Autonomous Prefecture, Deyang and Meishan in Sichuan Province.

In January 2019, *Sichuan Province Annals: Giant Panda* was published by Publishing House of Local Records. In June, *Giant Panda Pictorial Annals* came out which had been compiled jointly by the Editorial Board of Sichuan Province Local Records and Sichuan Province Forestry and Grassland Administration.

On March 8, 2019, was the opening day of the second session of the thirteenth National People's Congress. Peng Qinghua, Secretary of Sichuan Provincial CPC Committee and Deputy to the National People's Congress, holding a panda-shaped refrigerator magnet, introduced giant pandas to the audience.

On March 10, 2019, the plenary meeting of Sichuan delegation at the Second Session of 13th National People's Congress approved 13 proposals by the Sichuan delegation, including the optimization of nature reserves and protection of permanent basic farmland ranges, and more favorable policies and financial support for the establishment of giant panda national parks.

保护的意义在于和谐共生，人工繁育只是人类抢救大熊猫的第一步，最终，它们将被放归山林，重返自然（2017年） 周孟棋 摄

The significance of conservation lies in harmonious coexistence.Artificial breeding is only the first step for man to aid giant pandas.Eventually, they will be reintroduced to nature (2017). Photo by Zhou Mengqi

水墨画《生生不息》全卷，68cm×10600

Ink and wash: *Circle of Life* (whole), 68cm×10600cm (2015). By Tong Changxin.

2013 年，童昌信正在创作长卷《生生不息》 四川天娇大熊猫画院 供图

Tong Changxin is working on the 100-meter-long ink and wash painting *Circle of Life* in 2013.
Photo courtesy of Sichuan Tianjiao Giant Panda Art Academy

大熊猫栖息地水墨画百米长卷《生生不息》

2009—2015 年，巴蜀画派实力派人物童昌信耗时 7 年创作完成大熊猫栖息地水墨画百米长卷《生生不息》。《生生不息》卷长 106 米，高 0.68 米，共分为 4 卷，分别是：从夹金山至碧峰峡，从碧峰峡至天台山，从天台山至汶川三江源，从汶川三江源至九寨沟。163 只形态各异的大熊猫融入画卷，是传播大熊猫文化之巨作。

Circle of Life, a 100-meter-long ink and wash painting of the giant panda habitat

Tong Changxin, a luminary of Bashu Painting School, spent 7 years (2009-2015) creating this painting. As a masterpiece of panda culture, it includes 163 giant pandas in different looks and measures 106m in length and 0.68m in width; it consists of four sections such as from Jiajin Mountain to Bifengxia Gorge, from Bifengxia Gorge to Tiantai Mountain, from Tiantai Mountain to Three-River Source in Wenchuan and from Three-River Source to Jiuzhaigou.

大熊猫群像（2019 年） 周孟棋 摄

A mosaic giant panda of giant pandas (2019). Photos by Zhou Mengqi

剪纸《国之瑰宝 天之骄子》，直径 78cm（2013 年） 袁成祥 作品
Paper-cutting: *Treasure of China*, 78cm in diameter (2013). By Yuan Chengxiang

第一篇
物种

漆器《大熊猫》，直径 25cm（2005 年） 宋西平 作品

Lacquerwork: *The Giant Panda*, 25cm in diameter (2005). By Song Xiping

Part 1
Species

穿越时空，我们与大熊猫相遇

当人类尚处在猿人时代，

大熊猫便已经繁衍为一个空前兴旺的种族。

从出现至今，大熊猫历经大约 800 万年的进化，

是生物进化史上最著名的“活化石”，

是真正的时空使者。

The Giant panda has travelled to us all the way from remote antiquity

When humans were still in the *Pithecanthropus* age,
the giant panda had already proliferated into a flourishing species.
Since their emergence on earth,
they have gone through eight million years of evolution,
and earned the title of “a living fossil”,
a witness to whatever ecological changes on this planet.

大熊猫缓缓走来（2007年） 周孟棋 摄
A giant panda at his leisurely gait (2007). Photo by Zhou Mengqi

第一章 起源

千百万年对于古老的地球，不过转瞬即逝，对物种而言，却意味着严峻考验。各种动物为了生存，不断进化。

大熊猫的进化路径与众不同。今天的大熊猫主食是竹子，而它的肠道却依然保留着肉食动物的一些特征。它们脑容量一直很小，与几十万年前相差无几。

气候变化，成千上万的物种灭绝，大熊猫却活了下来。如果我们要用一个词来描述大熊猫的进化历程，没有什么比“奇迹”这个词更准确了。

大熊猫究竟是如何活下来的呢？人类依然在寻找答案。追溯它的起源，我们必须把时间向前推移约 2600 万年。一个古食肉动物的祖先类群，分化成早期的似熊类和古浣熊类。在距今 900 万～ 800 万年时，似熊类动物分化出始熊猫，而始熊猫（*Ailurarctos*）的主支则演化为大熊猫类。其中一种在距今 250 万年时出现，被称为小种大熊猫（*Ailuropoda microta*），并逐渐由食肉类动物演化为以竹子为主食的动物。

到了距今 70 万～ 10 万年前，气候出现多次冷暖交替，巴氏大熊猫（*Ailuropoda melanoleuca baconi*）逐渐繁衍兴盛。从已发现的大熊猫类化石来看，80% 以上都属于巴氏大熊猫。这种大熊猫亚种的体型大约比小种大熊猫大一倍，是由部分小种大熊猫在新的条件下逐渐进化而成。

当历史走到距今约 1 万年时，大熊猫个体已经演化得比巴氏大熊猫亚种小约 1/9 ～ 1/8，其体型大小已与现生大熊猫相似，因此有学者把近万年以

国画《玉色来天地》
140cm×70cm（2019 年） 李兵 作品

hinese painting: *Beauty in the World*, 140cm×70cm (2019). By Li Bing

来的大熊猫视为现生种（*Ailuropoda melanoleuca*），也就是我们现在见到的大熊猫。

约 800 万年的漫长进化历史，让大熊猫成为动物界极为罕见的“活化石”。它不仅为我们带来了无数未解之谜，而且使我们窥见百万年前的地球一貌。

保存在大英博物馆的大熊猫骨架模式标本
引自黄万波、魏光飚著《大熊猫的起源》，科学出版社，2010 年

A giant panda's skeleton type specimen preserved in the British Museum. Cited from Huang Wanbo, Wei Guangbiao, *The Origin of Giant Panda*, Science Press, 2010

Chapter 1 Origin

Millions of years is but a fleeting moment for the ancient earth. However, it means severe tests for all species and they had to evolve to survive.

The Giant panda has evolved distinctively from the other species. Nowadays, this species eats bamboo as its staple food but with the intestines resembling the characteristics of carnivores. It still has the same small brain as it did hundreds of thousands of years ago.

As a result of climate change, thousands of animal species went extinct, but the giant panda has survived. If we use one word to describe the evolution of this species, nothing is more appropriate than "miracle".

How has the giant panda managed to survive? Humans are still looking for an answer. To trace its origin, we must look back to about 26 million years ago when there existed a "taxon" of ancient ancestors of carnivores, which later differentiated into pseudo *Ursidae* and *Procyonidae*.9 million to 8 million years ago, Ursidae differentiated out into *Ailuaractos* (primal pandas), whose main lineage evolved into *Ailuropoda*. One lineage of *Ailuaractos*, named as *Ailuropoda microta* (pygmy giant panda), appeared 2.5 million years ago and gradually evolved from the meat-eating animal to the one feeding mainly on bamboo.

From 700,000 to 100,000 years ago, when there were many cycles of alternating hot and cold climate, *Ailuropoda microta* evolved into *Ailuropoda melanoleuca baconi*, was twice the size of the former. More than 80% of the giant panda fossils belong to this sub-species.

About 10,000 years ago, the figure of *Ailuropoda* was about 1/9-1/8 smaller than that of *Ailuropoda melanoleuca baconi*, and its body size was similar to that of today's giant panda, so some scholars regard the

《熊猫》，引自吴作人、萧淑芳《吴作人萧淑芳画选》，朝华出版社，1985 年

The Giant Panda. Cited from Wu uoren and Xiao Shufang, *Selected Paintings of Wu Zuoren and Xiao Shufang*, Blossom Press, 1985

species of giant panda that has lived in the recent 10,000 years as *Ailuropoda melanoleuca*, namely, the giant panda we see nowadays.

The evolution of eight million years makes the giant panda an extremely rare "living fossil" that carries mysteries too countless to be resolved. A study of this species can shed some light on what the earth was like millions of years ago.

绢本设色《熊猫》，69cm×69cm（2013 年） 吴越 作品
Ink and color on silk: *The Giant Panda*, 69cm×69cm (2013). By Wu Yue

大熊猫演化示意图
Graphic of Giant Panda Evolution

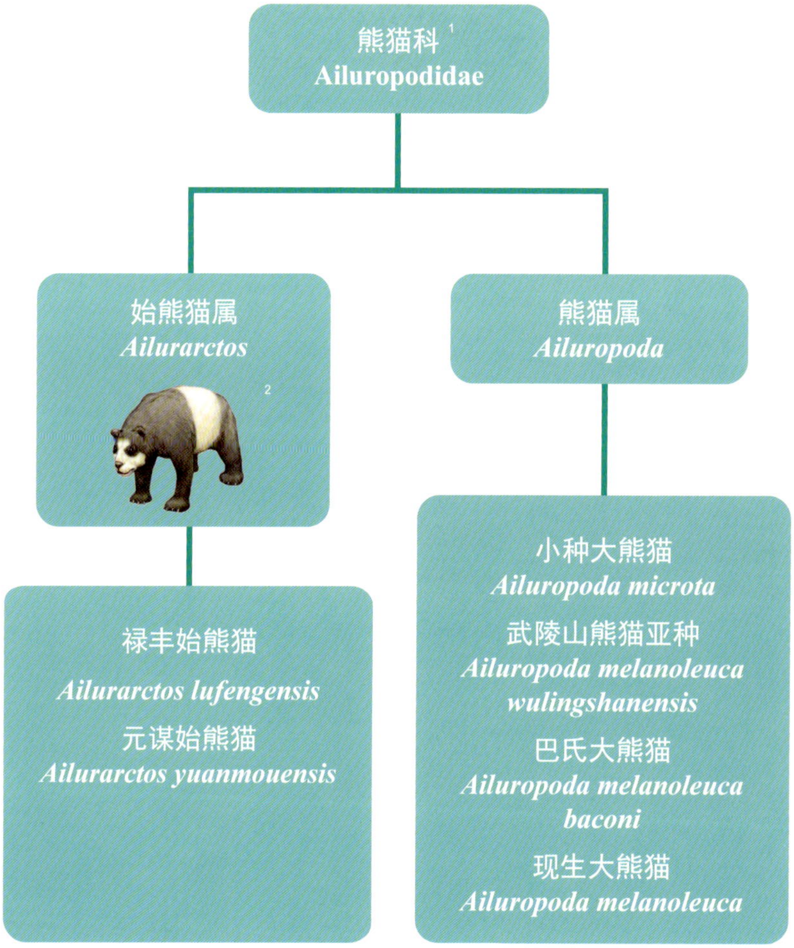

大熊猫演化示意图（2019 年） 考拉看看 制图 资料来源：黄万波、魏光飚著《大熊猫的起源》，科学出版社，2010 年

[1] 现已确定大熊猫为“熊科”。

[2] 始熊猫复原图 引自四川省档案局、四川省档案学校编《档案见证·大熊猫》，四川人民出版社，2018 年

Diagram of the Giant Panda Evolution (2019). By Koalacan. Cited from Huang Wanbo, Wei Guangbiao, *The Origin of Giant Panda*, Science Press, 2010

[1] Now the giant panda has been identified as *Ursidae.*

[2] A restoration drawing of *Ailuaractos*. Cited from *Archives Witness: The Giant Panda* compiled by Sichuan Province Archives Bureau and Sichuan Archives College, and published by Sichuan People’s Publishing House, 2018

始熊猫复原图
引自黄万波、魏光飚著《大熊猫的起源》，科学出版社，2010 年
A restoration drawing of *Ailuaractos*. Cited from Huang Wanbo, Wei Guangbiao, *The Origin of Giant Panda*, Science Press, 2010

始熊猫化石的发现

1978 年以来，中国科学院的科研人员在云南禄丰和元谋盆地发现了新的熊猫类化石，后被认为是始熊猫属的两个新种，分别被命名为禄丰始熊猫（*Ailurarctos lufengensis*）和元谋始熊猫（*Ailurarctos yuanmouensis*）。

Discovery of *Ailurarctos* Fossils

Since 1978, researchers from CAS have discovered new giant panda fossils in Lufeng County and Yuanmou Basin of Yunnan Province, which were later regarded as two new types of *Ailurarctos*, and named *Ailurarctos lufengensis* and *Ailurarctos yuanmouensis* respectively.

小种大熊猫 引自四川省档案局、四川省档案学校编《档案见证·大熊猫》，四川人民出版社，2018 年

Ailuropoda microta. Cited from *Archives Witness: The Giant Panda* compiled by Sichuan Province Archives Bureau and Sichuan Archives College, and published by Sichuan People's Publishing House, 2018

小种大熊猫

1962 年，在中国广西柳城巨猿洞内，中国科学院古脊椎动物与古人类研究所研究员裴文中[1]发现了一种大熊猫化石。在随后的研究中根据其个子矮小等特征而另立新种，即小种大熊猫 (*Ailuropoda microta*)。研究认为，小种大熊猫就是现生大熊猫的祖先。

Ailuropoda microta

In 1962, Pei Wenzhong,[1] a researcher from the Institute of Vertebrate Paleontology and Paleoanthropology, Chinese Academy of Science (CAS), discovered the fossils of a certain type of giant panda in a cave where the fossils of *Gigantopithecus* (giant apes) were discovered in Liucheng County, Guangxi, China. In subsequent studies, a new type of giant panda—*Ailuropoda microta* (pygmy giant panda)—was established according to such characteristics as short stature. Scholars suggest that this is the earliest ancestor of today's giant panda.

[1] 裴文中（1904. 1. 19—1982. 9. 18），河北丰南人，史前考古学家、古生物学家。

[1]Pei Wenzhong (January 19, 1904-September 18, 1982), a native of Fengnan County, Hebei Province, prehistoric archaeologist and paleontologist.

更新世小种大熊猫物种群 引自四川省林业厅编《图说大熊猫》，天地出版社，2015 年

Ailuropoda microta in Pleistocene. Cited from *The Giant Panda Pictures* compiled by Forestry Department of Sichuan Province and published by Tiandi Press, 2015

武陵山熊猫亚种

1978 年，中国科学院古脊椎动物与古人类研究所古生物学家王令红等研究员找到了一种新的大熊猫化石，后被命名为武陵山熊猫亚种（*A. M. wulingshanensis*）。

A. M. wulingshanensis

In 1978, Wang Linghong and other researchers from the Institute of Vertebrate Paleontology and Paleoanthropology of CAS found a new type of giant panda fossil, which was later named as *A. M. wulingshanensis.*

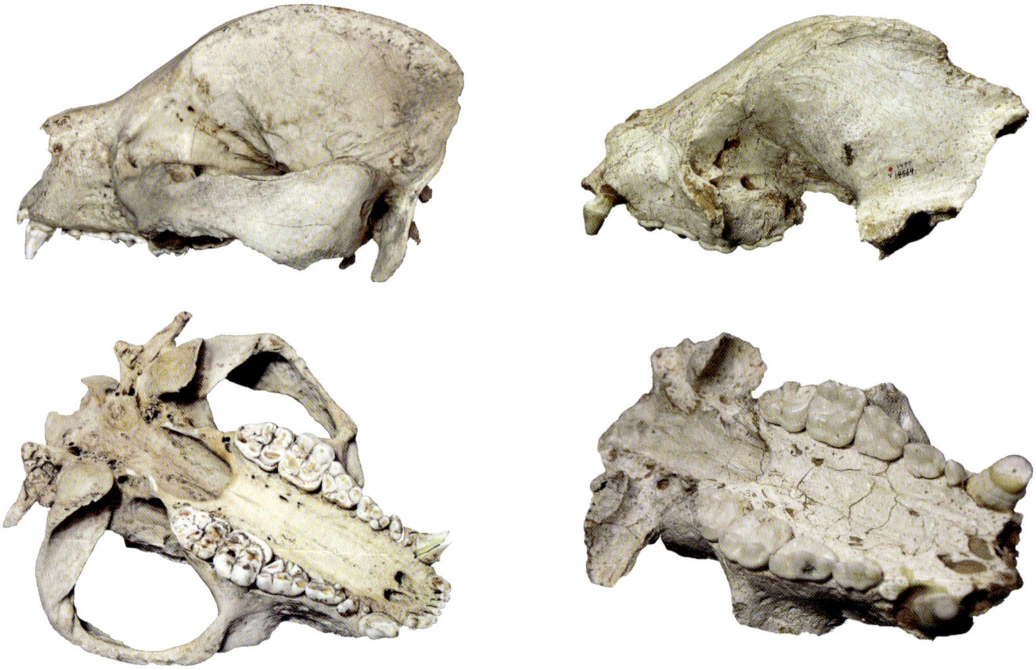

现生大熊猫头骨（左）和小种大熊猫头骨化石（右）的对比 引自魏辅文著《野生大熊猫科学探秘》，科学出版社，2018 年

A comparison of skull fossils of today's giant pandas (left) and those of *Ailuropoda micrta* (right). Cited from Wei Fuwen, *Scientific Exploration of Wild Giant Pandas*, Science Press, 2018

大熊猫头骨测量一览表
Table of Giant Panda Skull Measurements

单位：毫米 Unit：mm

指标（Item）	小种大熊猫 *Ailuropoda microta*		巴氏大熊猫 *A.m.baconi*		现生大熊猫 *A.melanoleuca*	
	n	x	n	x	n	x
颅基长 Condylobasal Length		—	2	267.5	47	255.5
颧宽 Zygomatic Width		—	2	222.5	43	210.9
顶骨部头宽 Head Width		—	2	113.5	18	99.4
眼眶后前额宽度 Forehead Width		—	3	61.5	35	52.5

说明：n：表示标本数；x：表示标本平均值

Illustration: n represents the number of specimens; x represents the average of specimens.

大熊猫头骨测量一览表（2019 年） 考拉看看 制图

资料来源：四川省地方志编纂委员会编《四川省志 · 大熊猫志》，方志出版社，2018 年

Table of measurements of the giant panda skull (2019). By Koalacan.

Cited from *Sichuan Province Annals: Giant Panda* compiled by Sichuan Local Records Editorial Committee, and published by Publishing House of Local Records, 2018

化石分布区

在世界范围内，大熊猫化石分布在中国、越南、老挝、缅甸和泰国。中国境内主要分布在陕西、甘肃、山西、福建、广东、广西、湖北、湖南、河南、四川、重庆、贵州、云南等地。

Fossil Distribution

Fossils of the giant panda were found in China, Vietnam, Laos, Myanmar and Thailand. In China this species is mainly distributed in Shaanxi, Gansu, Shanxi, Fujian, Guangdong, Guangxi, Hubei, Hunan, Henan, Sichuan, Chongqing, Guizhou and Yunnan.

大熊猫化石在中国的分布图
Distribution of Giant Panda Fossils in China

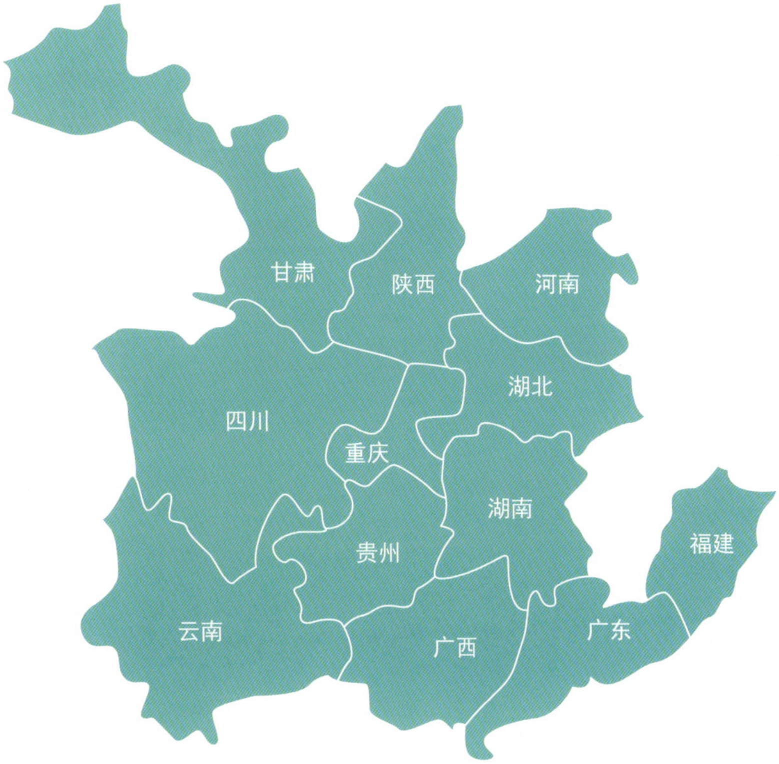

大熊猫化石在中国的分布示意图（2019 年） 考拉看看 制图
资料来源：四川省地方志编纂委员会编《四川省志·大熊猫志》，方志出版社，2018 年

Distribution of Giant Panda Fossils in China (2019). By Koalacan.

Cited from *Sichuan Province Annals: Giant Panda* compiled by Sichuan Local Records Editorial Committee, and published by Publishing House of Local Records, 2018

体型变化

如果将古今大熊猫的体型做一个对比图，就能明显观察到它体型的一些变化——这是一个由小到大再到小的过程。魏辅文、胡锦矗等学者用野外收集到的大熊猫头骨数据与现有的巴氏大熊猫化石数据进行对比，发现巴氏大熊猫的头骨和牙齿的平均值都明显大于现生大熊猫。[1]

Change in Body Size

If we look at the giant panda across different evolutionary stages, we can easily discern its change in body size: it grew bigger and then smaller. Scientists like Wei Fuwen and Hu Jinchu compared the data of giant panda skulls collected in the wild with those of the fossils of *Ailuropoda melanoleuca baconi*, and found that the average sizes of the skulls and teeth of *Ailuropoda melanoleuca baconi* were significantly greater than those of *Ailuropoda melanoleuca* today.[1]

[1] 魏辅文、吴毅、袁重桂、胡锦矗、钟肇敏．大熊猫体型变化及兴衰分析 [J]．四川师范学院学报（自然科学版），1990, 11 (1):pp. 23—28.

[1] Wei Fuwen, Wu Yi, Yuan Chonggui, Hu Jinling and Zhong Zhaomin. *Analysis of the Change of Giant Pandas' Body Shape and Their Flourishing and Their Boom and Bust* [J]. Journal of Sichuan Teachers' College (Natural Science Edition), 1990, 11 (1): pp. 23-28.

数码插画《熊猫进化图》，60cm×90cm（2019 年） 殷頔 作品
Digital illustration: *Evolution of the Giant Panda*, 60cm×90cm (2019). By Yin Di

第二章 古名

在中国古籍中，记载有大量奇珍异兽，其中不少被认为与大熊猫的外形、习性相似。如最常出现的两个名字为“貔貅（pí xiū）”和“貘（mò）”。

1956年7月6日，周建人[1]首次提出大熊猫古名的研究考据。他在《人民日报》和中国作家协会选编的《散文小品选》上刊发了一篇散文，名为《关于大熊猫》。文章从到北京西郊动物园观看大熊猫展出，引申出大熊猫的古名，其意在向读者讲述和普及生物学知识。文章写道：“古时候所谓‘貔貅’，大概便是指这种动物。”

1980年后，大熊猫的古名出现了更多的提法和考证。除“貔貅”“貘”“驺（zōu）虞”外，还有“貊（mò）”“白罴（pí）”“貘白豹”等多种说法。

2008年8月，四川省旅游局巡视员孙前、中国科学院动物研究所研究员何芬奇、四川大学研究生孙震在《动物分类学报》（2008年第4期）上发表《大熊猫古名研究》，他们将“貔貅”“貘”“驺虞”这三种说法的古籍原文全部引用出来，并进行分析考证，最后得出结论，三种称谓都不能肯定是大熊猫的古名。

胡锦矗、夏勒等曾经提出上面三种称谓是大熊猫的古名，但他们在《卧龙的大熊猫》一书中也指出：“（古代自然历史的）文字记载都是基于经传的描述或注疏，而不是以实际观察为根据。这就引起了名称和记载上的混淆，这些名称和记载可能是也可能不是指熊猫。”[2]

生活在中国大熊猫栖息地的群众则依据大熊猫的体色、形貌、食性等特征，称呼它为“竹熊”“花熊”“白熊”“花白熊”“食铁兽”“大浣熊”等。

国画《中华国宝 熊猫呈祥》
68cm×137cm（2007年）
邓康林 作品

[1] 周建人（1888. 11. 12—1984. 7. 29），浙江绍兴人，鲁迅与周作人的胞弟，著名社会活动家、生物学家。

[2] 胡锦矗、夏勒等．卧龙的大熊猫[M]．成都：四川科学技术出版社，1985.

Chapter 2 Archaic Names

Chinese painting: *Auspicious Giant Pandas: Chinese National Treasure*, 68cm×137cm (2007). By Deng Kanglin

In ancient Chinese books, there are records of numerous rare animals, many of which are thought to have similar appearances and habits of the giant panda. Among the various names, *pixiu* and *mo* are the commonest.

On July 6, 1956, Zhou Jianren[1] initiated the exegesis for the names of the giant panda in history. He published an essay titled "About the Giant Panda" in *The People's Daily* (which was included in *A Collection of Essays and Short Sketches* edited by the Chinese Writers Association). The essay begins with his visit to Beijing Zoo to see the giant panda, followed by an introduction to their names in history, aiming to popularize the biological lore. The article notes: "In ancient times the so-called *pixiu* probably refers to the giant panda."

After 1980, more exegeses were made for the ancient names of the giant panda. Added to *pixiu*, *mo* and *zouyu* are moo, *baipi* and *mo* white panther.

In August 2008, Sun Qian (consultant to Tourism Administration of Sichuan Province and famous scholar of giant panda culture), He Fenqi (researcher at Institute of Zoology of CAS), and Sun Zhen (graduate student of Sichuan University) co-authored an article titled "Annotation on the Ancient Chinese Names of the Giant Panda" and published in the magazine *Zoological Systematics* (Vol. 4, 2008). They studied all the ancient books which include *pixiu*, *mo* and *zouyu*, and finally came to doubt whether these three terms were the archaic names for the giant panda.

Hu Jinchu and George B. Schaller once suggested that the three terms above were the names for the giant panda in ancient times, but they also pointed out in their co-authored book *The Giant Panda of Wolong*, "... as is generally true of ancient natural history in any culture, writings were based on the descriptions and inferences from the classics, not just on actual observations. This resulted in confusing names and descriptions that may or may not refer to the panda."[2]

People in the habitat for the giant panda call it bamboo bear, banded-bear, white bear, banded white bear, iron-eating animal and big raccoon according to its color, appearance and diet.

[1]Zhou Jianren (November 12, 1888-July 29, 1984), a native of Shaoxing City, Zhejiang Province, brother of Lu Xun and Zhou Zuoren, a social activist and biologist.

[2]Hu Jinchu, George B. Schaller, et al., *The Giant Panda of Wolong* [M]. Chicago and London: The University of Chicago Press, 1985.

貔貅

《诗经 · 大雅 · 韩奕》中记载了“貔貅”这一动物：“献其貔皮，赤豹黄罴。”形似大熊猫的动物图案，则见于3000多年前出现的貔貅旗。《礼记 · 曲礼上》也有记载：“前有挚兽，则载貔貅。”但是貔貅是否真为大熊猫，有待考证。

Pixiu

Both *Shi Jing* (*Book of Songs*) and *Li Ji* (*Book of Rites*) bear the name of the mythical beast *pixiu*, but nothing else is mentioned in either. A pattern featuring a *pixiu* that looked like the giant panda, appeared on a flag more than 3,000 years ago. However, whether *pixiu* refers to the giant panda remains to be verified.

貘

《本草纲目》记载：“貘膏性利，铜、铁、瓦器盛之悉透，惟以骨盛则不漏。”有学者研究，古文献中所说的“貘”，是至今存在于东南亚一带的马来貘，而非大熊猫。

Mo

According to *Compendium of Materia Medica*, “*Mo* oil can penetrate copperware, ironware and crockery, but not the bone.” But some scholars proposed that *mo* mentioned in ancient literature is a Malay tapir that still exists in Southeast Asia, rather than the giant panda.

驺虞

《资治通鉴》中有一段关于“驺虞旗”的记载：“驺虞仁兽，故以解兵。”“驺虞旗”代表和平友好，两军交战，只要一方举起“驺虞旗”，即表示要求停止战争。

Zouyu

A history book named *Zizhi Tongjian* (*History as a Mirror*) records a flag featuring a *Zouyu*, which represented peace and friendship. When two armies were locked in war, one of them could raise the “*Zouyu* Flag” as a sign of truce.

* 篆书（2019年） 马睿 作品

白熊

《毛诗草木鸟兽虫鱼疏》之注云："貔似虎或曰似熊，辽东人谓之白熊。" 有人认为 "白熊" 即是大熊猫。但据考证历史上大熊猫并未分布在辽东。

White Bear

A white bear is recorded in *Introduction to Animals and Plants*, an exegesis for the animals and plants mentioned in *Book of Songs*, "*Pixiu* looks like a tiger or a bear, so locals east of Liaohe River call it a white bear." Some people believe this "white bear" is one of the giant panda which, however, has never been found east of the River.

貘白豹

《尔雅 · 释兽》称一种动物为 "貘白豹"。有学者认为它是大熊猫的古名之一，但也有学者认为是指雪豹。

Mo White Panther

Mo white panther is recorded in *Er Ya*. Some scholars believe it to be one of the ancient names for the giant panda, while others propose that it refers to the snow leopard.

食铁兽

《尔雅 · 释兽》一书的注中写道："似熊，小头，卑脚，黑白驳，能舐食铜铁及竹骨。骨节强直，中实少髓，皮辟湿。" 这与现生大熊猫的形态和习性十分相近。

Iron-eating Beast

Er Ya also notes that this beast "resembles the bear, with a small head, small feet, and black-and-white spots in its fur, capable of devouring copper, iron and bamboo. It looks very strong, and its thick fur helps keep out the wet weather." This resembles the appearance and habit of the modern giant panda.

*Seal characters of the giant panda's archaic names (2019). By Ma Rui

明万历年间的大熊猫形象（驺虞）*

The drawing of the giant panda in the Wanli reign of Ming Dynasty (*Zouyu*)

明弘治年间的大熊猫形象（貘）*

The drawing of the giant panda in the Hongzhi reign of Ming Dynasty (*Mo*)

猛豹，明万历年间（左）、日本江户时代（中）、清光绪年间（右）的大熊猫形象*

The drawings of the giant panda—*Mengbao*—in the Wanli reign of Ming Dynasty (left), in the Edo age of Japan (middle), and in the Guangxu reign of Qing Dynasty (right)

*引自赵学敏编《大熊猫——人类共有的自然遗产》，中国林业出版社，2006 年

*Cited from Zhao Xuemin, *The Giant Panda: Natural Heritage of Humanity*, China Forestry Publishing House, 2006

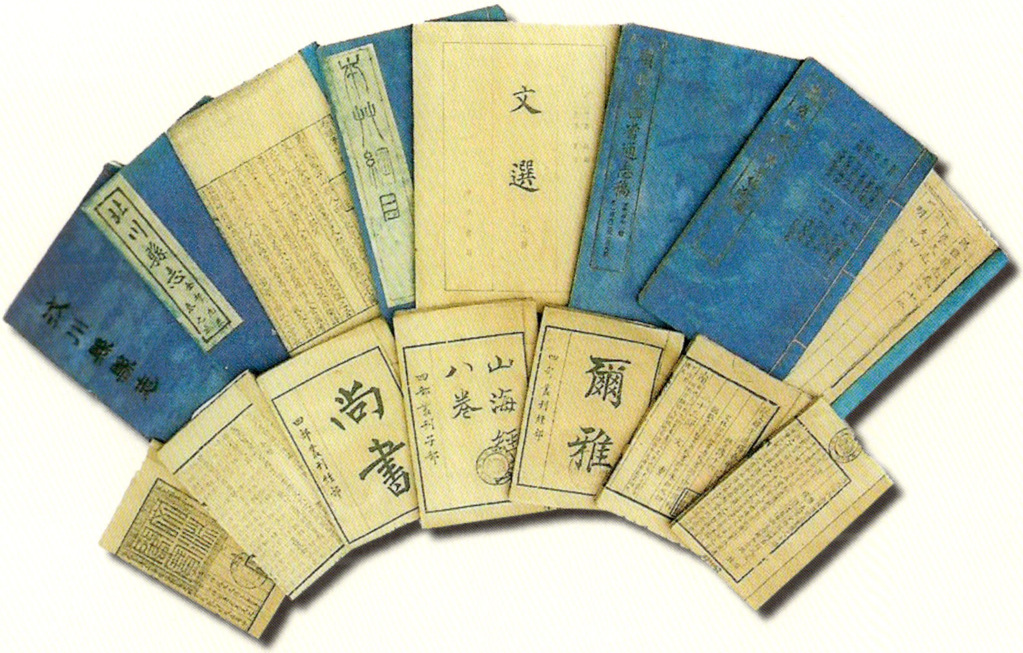

《山海经》《尚书》《尔雅》等古籍都对大熊猫进行了记述
引自赵学敏编《大熊猫——人类共有的自然遗产》，中国林业出版社，2006 年
Reports about the giant panda in *Shanhai Jing* (*The Classic of Mountains and Rivers*), *Shangshu* (*The Book of Historical Documents*), *Er Ya* (the first dictionary in China), etc. Cited from Zhao Xuemin, *The Giant Panda: Natural Heritage of Humanity*, China Forestry Publishing House, 2006

《史记·五帝本纪》有黄帝驯养虎、豹、熊和貔貅助战的记载 成都大熊猫繁育研究基地 供图
Reports in *Shi Ji* (*Records of the Historian*) that Huangdi (Yellow Emperor) raised and trained beasts like tigers, leopards, bears and *pixiu* for wars. Photo courtesy of Chengdu Research Base of Giant Panda Breeding

第三章　相遇

尽管有关大熊猫的古名依然有争议，但是历史上中华民族与大熊猫的接触却一直存在。比如在汉文帝的母亲薄太后[1]的陵墓陪葬坑里就发现了大熊猫的骨骸。

大熊猫走入世界视野，则是在清同治八年（1869年）。

清同治八年（1869年）2月28日，法国博物学家阿尔芒·戴维神父到四川省穆坪（今宝兴县）邓池沟天主教堂，在当地采集各种动植物标本。3月11日，一位李姓猎户邀请戴维到家中做客。戴维看到墙上有一张黑白毛皮。对动植物极有研究的戴维发现在自己的记忆中从未见过这种动物，包括他读过的任何一本动物学专著中，都没有关于这种动物的记载。于是戴维请当地猎户为他捕捉这种动物。

戴维在他日记中写道："我终于看到了一张久闻的黑白熊皮，而且它相当巨大。它真是个奇妙的生物。当猎人告诉我，我不久之后也能拥有它时，我简直喜出望外。他们说明天就要进山去打猎，这将会是一件全新而有趣的科学发现。"[2]

3月23日，猎人们为戴维找来1只大熊猫幼崽，但不幸死亡，戴维虽感痛惜，但仍然仔细观察，认为这是"熊类中的一个新种"。4月1日，1只成年活体大熊猫被猎人们捕到。戴维见到大熊猫，如获至宝。

在充分观察之后，戴维将大熊猫的皮毛和骨骼分别进行整理，对骨骼作了编号记录，装箱运回法国。和大熊猫标本一起被运回去的还有戴维写的论文，他希望巴黎自然历史博物馆主任米勒·爱德华收到标本后，邀请相关方面专家进行深入研讨，然后公布对这种物种的描述。

[1] 即薄姬（公元前215年—公元前155年），汉高祖刘邦的嫔妃、汉文帝刘恒的生母。

[2] 阿尔芒·戴维. 戴维日记[M]. 海伦·佛克斯，译. 波士顿：哈佛大学出版社，1949:p276.

油画《打闹》，120cm×180cm（2018 年） 万山河 作品
Oil painting: *Frolicking*, 120cm×180cm (2018). By Wan Shanhe

水彩画《花与大熊猫》，53cm×38cm（2019 年）
德阳市罗江区第一幼儿园 田羽轩 作品
指导老师 黄建华
Watercolor painting: *Flowers and Giant Pandas*, 53cm×38cm (2019). By Tian Yuxuan from No.1 Kindergarten in Luojiang District, Deyang City, under the instruction of Huang Jianhua

爱德华收到戴维的文章和标本后，在1869年出版的《巴黎自然历史博物馆之新档案》第5卷第13页上发表了戴维对大熊猫的描述，并将标本重新整理后向公众展示。这吸引了众多参观者，也轰动了欧洲生物学界。戴维因此成了第一个向世界介绍中国大熊猫的外国人，他所发现和描述的“黑白熊”，成为记载和定名大熊猫这一新物种的模式标本[1]。

[1] 模式标本即作为规定的典型标本，在确定及发表某一群生物的学名时，应指出此学名的特征与作为分类概念标准的模式标本，但并不一定限于此群的典型代表。以前人们并未重视典型这个观念，因此造成许多标本未能保存下来。

Chapter 3 Discovery

Although the ancient names for the giant panda remain controversial, man has had associations with this species since ancient times. For example, bones of the giant panda were found in the mausoleum of Empress Dowager Bo[1], mother to Emperor Wen of the Han Dynasty.

But it was not until 1869 that the giant panda appeared on the world stage.

On February 28, 1869, French naturalist Fr Jean Pierre Armand David came from Beijing to Dengchigou Catholic Church in Muping (now Baoxing) County, Sichuan, to collect specimens of various plants and animals. On March 11 when invited home by a hunter surnamed Li, David caught a glimpse of a black-and-white pelt on the wall. Steeped in zoology as he was, he found that he had never seen such an animal before, and there was no report of it in any zoological treatise he had ever read. So he asked Li to capture one for him.

David noted in his diary: "I see a fine skin of the famous white and black bear (*Ailuropoda melanoleucus*), which appears to be fairly large. It is a remarkable species and I am delighted when I hear my hunters say that I shall certainly obtain the animal within a short time. They tell me they will go out tomorrow to kill this animal, which will provide an interesting novelty for science."[2]

On March 23, the hunters found a panda cub for David, but unfortunately, it was a dead one. Although David felt sorry for it, he still made a careful observation and decided that it was "a new species of bear." On April 1, a live adult panda was captured by hunters and David was so pleased with that.

It was a pity that this adult panda also died. David took their furs and bones (which he got numbered) and shipped them back to France. Along with the specimens was a paper written by him. The recipient of these treasures was Milne Edwards, director of the National Museum of Natural History in Paris. He appealed to Edwards that experts be asked to make studies of these specimens, and that their findings be published.

After receiving David's paper and specimens, Edwards published David's description of this species on page 13, volume 5 of *New*

布面油画《地中海的视觉日志》
120cm×200cm（2007 年）
张奇开 作品

[1]Bo (215 BC-155 BC), a concubine of Liu Bang, Emperor Gaozu of the Han Dynasty; and mother of Liu Heng, Emperor Wen of the Han Dynasty.

[2]Armand David, *Abbe David's Diary* [M]. Trans. Helen M. Fox. Boston: Harvard University Press, 1949: p276.

Oil painting on canvas: *Mediterranean Visual Log*, 120cm×200cm (2007). By Zhang Qikai

1938 年，大熊猫“潘多拉”在华西协合大学内与孩子们嬉戏 谭楷 供图

The giant panda Pandora was frolicking with children at West China Union University in 1938. Photo courtesy of Tan Kai

Archives of National Museum of Natural History in Paris released in 1869. The specimens were reclassified and made available to the public, attracting a great number of visitors and arousing a sensation in the European biology circles. David thus became the first scientist outside of China to introduce the Chinese giant panda to the world. The “black-and-white bear” he found and described was taken as the type specimen[1] of the giant panda.

[1]The type specimen is the prescribed typical specimen. When establishing and publishing the scientific name for a taxon, it is necessary to point out its features, and type specimen which is not necessarily the typical representative of the taxon. In the past, researchers did not pay much attention to the concept of type specimen, and that is the reason that many specimens have failed to be preserved.

阿尔芒·戴维，左图为 1872 年摄于北京，右图摄于 1862 年到中国之前 孙前 供图

The left photo was Armand David taken in Beijing in 1872, and the right one before he was coming to China in 1862. Photo courtesy of Sun Qian

阿尔芒·戴维

阿尔芒·戴维（1826.9.7—1900.11.10），法国西部巴斯克人，博物学家，在中国进行了 12 年的博物收集、研究工作。清同治元年（1862 年）7 月 5 日，戴维来到北京，开启了他在中国的探险之旅。他在中国科学发现了大熊猫、麋鹿、川金丝猴、扭角羚、珙桐等 189 个新物种，并将这些新物种介绍给全世界。

Fr. Jean Pierre Armand David

Fr. Jean Pierre Armand David (September 7, 1826-November 10, 1900) was a Basque and naturalist from western France. He stayed in China for 12 years to collect samples for his research. On July 5, 1862, David arrived in Beijing and began his adventures in China. He discovered 189 new species in China, including the giant panda, David's deer, golden snub-nosed monkey, takin and dove tree, and introduced them to the world.

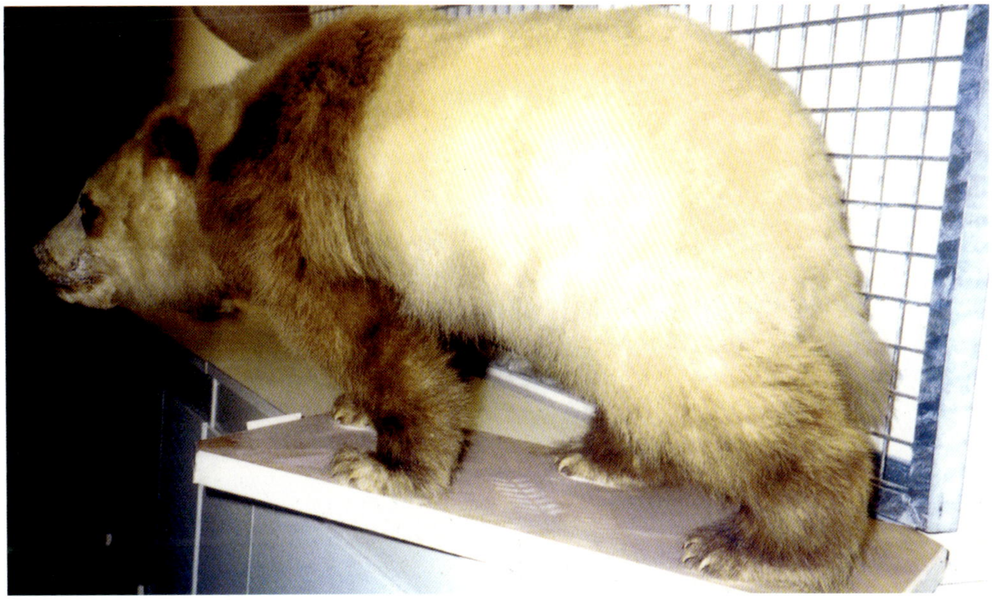

藏于巴黎自然历史博物馆的大熊猫标本（2002 年） 孙前 摄
The giant panda specimens in the National Museum of Natural History in Paris (2002). Photo by Sun Qian

走向世界

戴维将中国四川的大熊猫标本送往巴黎自然历史博物馆展出。大熊猫从此由中国走向世界，进入自然科学研究的范畴。中国四川省宝兴县成为第一个大熊猫模式标本的产地，全世界人民开始认识大熊猫这一新物种。

First Show on the World Stage

David sent specimens of the giant panda for display in the National Museum of Natural History in Paris. The exhibition meant that it had gone out of China and entered the scope of scientific study of natural science. Baoxing County in Sichuan turned out to be the locality from which the first type specimen of the giant panda was produced. People worldwide had thus come to know this new species.

邓池沟天主教堂全景（2015 年） 高华康 摄
Panorama of the Dengchigou Catholic Church (2015). Photo by Gao Huakang

四川省雅安市宝兴县邓池沟 引自四川省档案局、四川省档案学校编《档案见证 • 大熊猫》，四川人民出版社，2018 年
Dengchigou in Baoxing County, Ya’an City, Sichuan Province.Cited from *Archives Witness: The Giant Panda* compiled by Sichuan Province Archives Bureau and Sichuan Archives College, and published by Sichuan People’s Publishing House, 2018

邓池沟天主教堂内景（2019 年） 考拉看看 供图
The interior of the Dengchigou Catholic Church (2019). Photo courtesy of Koalacan

邓池沟天主教堂

位于中国四川省雅安市宝兴县盐井乡，清道光十九年（1839 年）由法国传教会修建。教堂的建筑风格独具匠心，是中西文化结合的产物。清同治八年（1869 年），戴维神父在这里见到了第一只活体大熊猫，并写下了鉴定意见连同标本一起寄回巴黎自然历史博物馆。

Dengchigou Catholic Church

Located in Yanjing Township, Baoxing County, Ya'an City, Sichuan Province, as a product of the combination of Chinese and western architectural styles, it was built by the French Congregation of the Mission in 1839. In 1869, Father David saw a live giant panda here and sent his appraisal and specimens back to the National Museum of Natural History in Paris.

《戴维日记》摘录

戴维在他的日记中详细记录了大熊猫的科学发现。

3 月 23 日 猎人们在进山十天后终于回来了。他们给我带回来了一只白熊幼崽。原本这只白熊他们抓到时是活着的，但不幸的是，他们要将它杀死，因为这样方便运输。他们以极高昂的价格将这只幼崽卖给我。而且它全身雪白，只有四肢、耳朵和眼睛周围是深黑色。这种颜色和我上次在猎人李家中看到的那张熊皮一模一样，那是一张成年熊的皮。这一定是熊类的新物种。它的特殊之处不仅在于它的毛色，还有它毛茸茸的小爪子以及其他特征。

4 月 1 日 他们给我带来一只白熊，并告诉我它已经成年。它的毛色和我原先买的那只幼崽十分相似，只不过它深色的部位不及幼崽黑色部位颜色更深，白色的毛发有些脏了。这只成年熊的头很大，口鼻部位圆圆的而且很短，而且和北京地区的熊有很大不同。[1]

《戴维日记》立体封面
引自孙前著《大熊猫文化笔记》，五洲传播出版社，2009 年
Abbe David's Diary (3D book cover). Cited from Sun Qian, *Notes of the Giant Panda Culture*, China Intercontinental Press, 2009

[1] 阿尔芒·戴维. 戴维日记 [M]. 海伦·佛克斯，译. 波士顿：哈佛大学出版社，1949: pp. 282—283.

Excerpts from *Abbe David's Diary*

David detailed the discovery of the giant panda in his diary.

March 23 My Christian hunters return today after a ten-day absence. They bring me a young white bear, which they took alive but unfortunately killed so it could be carried more easily. The young white bear, which they sell to me very dearly, is all white except for the legs, ears, and around the eyes, which are deep black. The colors are the same as those I saw in the skin of an adult bear the other day at the home of Li, the hunter. This must be a new species of *Ursus* (*Ailuropoda*), very remarkable not only because of its color, but also for its paws, which are hairy underneath, and for other characters.

April 1 They bring me a white bear (*Ailuropoda melanoleucus*) which they tell me is fully adult. Its colors are exactly like those of the young one I have, only the darker parts are less black and the white more soiled. The animal's head is very big, and the snout round and short instead of being pointed as it is in the Pekin bear.[1]

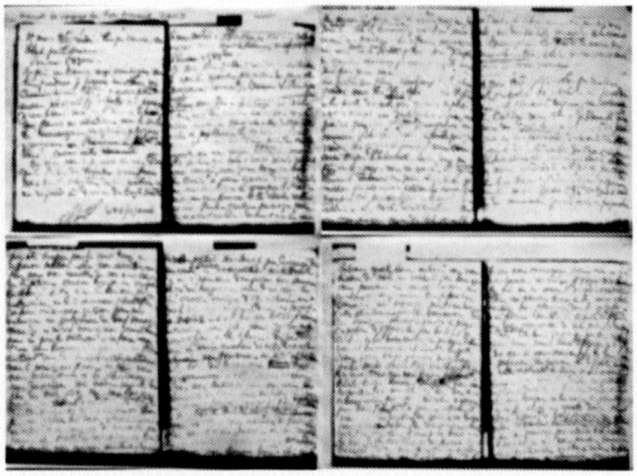

《戴维日记》复印件

引自孙前著《大熊猫文化笔记》，五洲传播出版社，2009 年

Photocopy of the manuscript of *Abbe David's Diary*.

Cited from Sun Qian, *Notes of the Giant Panda Culture*, China Intercontinental Press, 2009

[1] Armand David, *Abbe David's Diary* [M]. Trans. Helen M. Fox. Boston: Harvard University Press, 1949: pp. 282-283.

三兄弟（2008 年） 张志和 摄
Three brothers (2008). Photo by Zhang Zhihe

第四章 身份

清同治九年（1870 年），爱德华将大熊猫与中国西藏的其他动物进行对比分析，认为应该将大熊猫新立一属*Ailuropoda*（熊猫属）。因为大熊猫这一物种既不是熊，也不是猫，而是与中国西藏发现的小熊猫相似的另一种较大的“猫熊”，将其置于 *Ursus*（熊属）之下不合适。但是鉴于“*Ailuropoda*”一词已经被使用过，爱德华又于 1872 年将该属改为“*Ailuropus*”，即与小熊猫相似的“黑白熊”。

英国大英博物馆主任兰克斯特研究了保存于该馆的两具大熊猫标本（皮和骨骼），并于 1901 年发表论文，认为大熊猫接近于浣熊，尤其是与小熊猫更为接近，应该称为“硕大的熊猫”（Giant Panda）。后来，分类学家们仔细研究了大熊猫的学名，他们仍然采用爱德华最先定的名字，即 *Ailuropoda melanoleuca*，翻译成为“猫熊”（Cat bear）或“大猫熊”。

中文“熊猫”名称的由来，有着不同说法。比较流行的说法是：民国 28 年（1939 年）8 月 11 日，中国第一次在重庆北碚平民公园（今北碚公园）展出饲养的大熊猫。当时展出的标牌上分别用中、英文字书写“猫熊”的学名与中文名——上排从左至右用英文横写猫熊学名，下排为了和英语的书写方式保持一致，亦从左至右用中文写上“猫熊”两字。由于当时中文的读写习惯是从右至左，因此观众都将“猫熊”读成了“熊猫”。

但是有专家认为，上述说法是 1950 年后的讹传，当时展出的称呼是“白熊”。并且在民国 28 年（1939 年）之前，已经有部分文献和书籍开始使用“熊猫”这一词汇。“熊猫”一词来源的真实情况仍有待考证。

布面油画《我，中国最小的公民》
140cm×195cm（2010 年） 张奇开 作品

Oil painting on canvas: *I Am the Smallest Citizen of China*, 140cm×195cm (2010). By Zhang Qikai

Chapter 4 Identity

国画《熊猫乐园系列》，68cm×68cm （2017 年） 成都市龙江路小学
1. 徐浩然 作品 2. 李可怡 作品
指导老师 余文兰
Chinese painting: *Panda Paradise Series*, 68cm×68cm (2017).
By Xu Haoran and Li Keyi from Chengdu Longjiang Road Primary School, under the instruction of Yu Wenlan

In 1870, Edwards compared the giant panda with other animals in Tibet, China, and decided that a new genus of *Ailuropoda* should be set up. Because the panda is neither a bear nor a cat, but a larger "cat bear" like the red panda found in Tibet, so it would be inappropriate to be placed under *Ursus* (genus for the bear). However, since *Ailuropoda* had been used, Edwards changed it in 1872 to *Ailuropus*, which meant a black-and-white bear resembling the red panda.

Ray Lankester, director of the British Museum, studied the specimens (furs and bones) of two giant pandas kept in the museum, and published a paper in 1901. He believed that these specimens had close relationships to the raccoon family, especially. closer to the red panda, and that they should be named the Giant Panda. But taxonomists still use *Ailuropoda melanoleuca*, the name set by Edwards, which means "cat bear" or "giant cat bear."

There are different versions about the origin of the Chinese name "*xiong-mao*" (literally bearlike cat). The more popular one is: On August 11, 1939, live giant pandas were on display for the first time in Beibei Civilian Park (Beibei Park now) in Chongqing City. Their Latin and Chinese names were both written on a sign—the Latin above the Chinese. As Latin was written from left to right, so Chinese (*mao-xiong*) just followed suit, rather than in the conventional way (from right to left). An interesting thing occurred, the audience read it from right to left as "*xiong-mao*" (bearlike cat) instead of "*mao-xiong*" (catlike bear).

But some experts believe that the above account is but an invention. Actually, before 1939, some documents and books began to use the word "*xiong-mao*." Then in 1950, the panda at display was called "*bai-xiong*" (white bear). Thus, the controversy over the origin of the word "*xiong-mao*" is likely to go on.

米勒·爱德华鉴定文件中的大熊猫，是有史以来的第一张大熊猫画
引自孙前著《大熊猫文化笔记》，五洲传播出版社，2009 年
The giant panda in Milne Edwards' identification document, the first ever drawing of the giant panda. Cited from Sun Qian, *Notes of the Giant Panda Culture*, China Intercontinental Press, 2009

单独的属

1874 年，爱德华在《哺乳动物研究》上发表有关论文，再次论述了自己的观点：大熊猫应划为单独的属，这比把它归入食肉目中的任何一属都更有依据。

Independent Genus

In an 1874 paper in the journal of *Mammalian Studies*, Edwards reiterated his view that the giant panda should be classified as a separate genus, rather than put into any genus under carnivora.

分类争论

对于大熊猫的分类，动物学界一直存在着争议。过去不少学者将它列为独立的大熊猫科，但近年来国内外多数学者认为应将它列入熊科。

Controversy over Classification

Classification for the giant panda has always been controversial in zoology. In the past, many scholars catalogued it as an independent family, but in recent years, most scholars in and outside China propose that it should be included in the bear family.

1935 年出版的《中华大字典》对熊猫的描述
引自《中华大字典》，上海中华书局印行，1935 年
Description about the giant panda in The *Chinese Great Dictionary* published in 1935. Cited from *The Chinese Great Dictionary*, Shanghai Classics Publishing House, 1935

“熊猫”来源

民国 24 年（1935 年）出版的《中华大字典》里，“熊”词条释义有：“熊猫，兽名。似猫而善升木。”从附图看，是小熊猫。民国 27 年（1938 年）出版的《辞海》中，收录有“熊猫”词条，释义为：“怪兽名，产新疆……距今六十年前，为法国科学家比利大卫氏所发现。”后者则为大熊猫，但产地记载有误。

Origin of the name *xiong-mao*

The entry for *xiong* (bear) in the 1935 edition of The *Chinese Great Dictionary notes*, “*Xiong-mao* is the name of a beast, which looks like a cat and excels in climbing trees.” The illustration attached is a red panda. In the 1938 edition of the dictionary *Ci Hai* (meaning a sea of words), there is an entry of *xiong-mao*: “Name for a kind of monster that inhabits Xinjiang... discovered sixty years ago by the French scientist Jean Pierre Armand David.” The monster is the giant panda, but the dictionary made a mistake concerning its home.

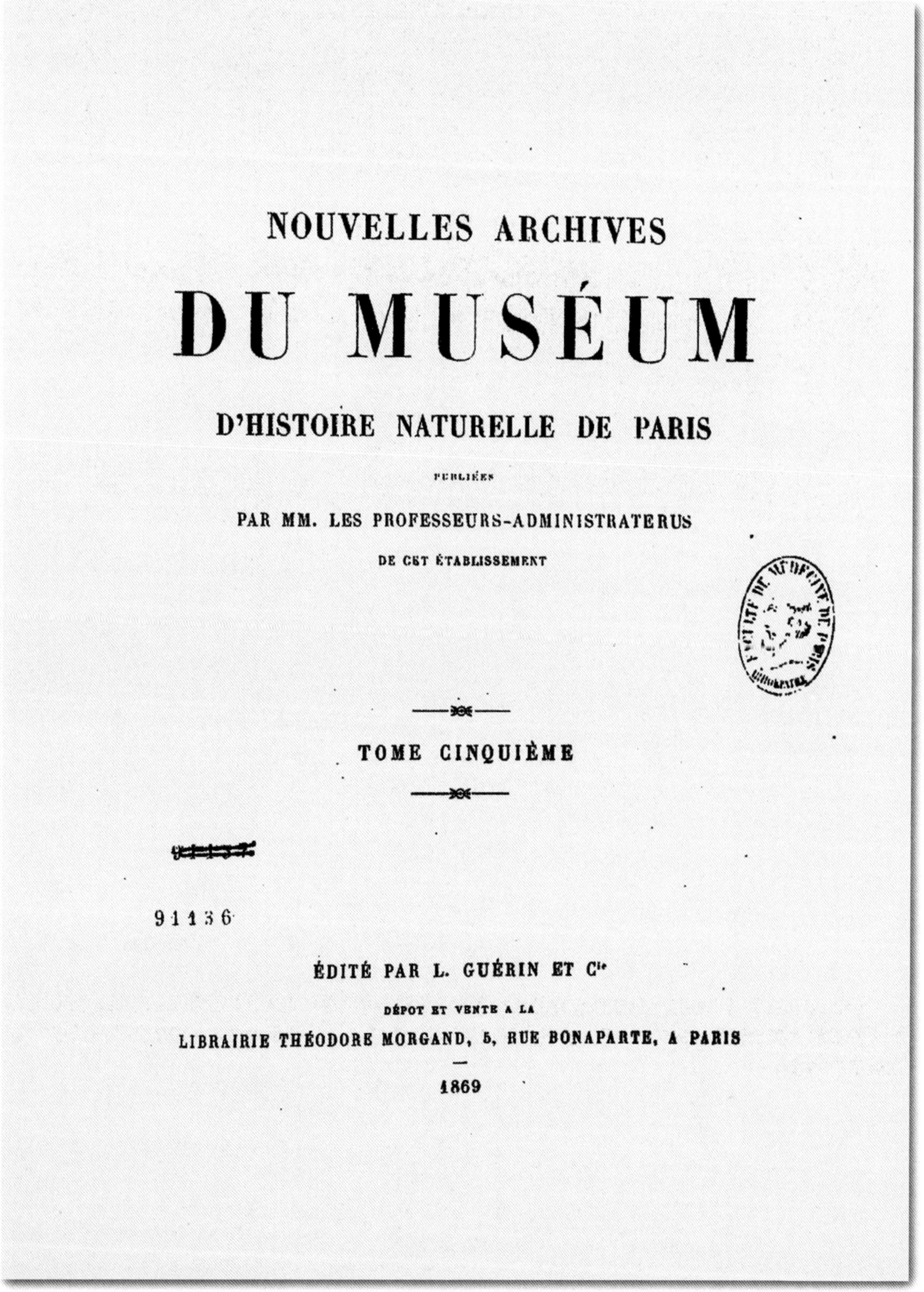

NOUVELLES ARCHIVES

DU MUSÉUM

D'HISTOIRE NATURELLE DE PARIS

PUBLIÉES

PAR MM. LES PROFESSEURS-ADMINISTRATERUS

DE CET ÉTABLISSEMENT

TOME CINQUIÈME

ÉDITÉ PAR L. GUÉRIN ET Cie

DÉPOT ET VENTE A LA

LIBRAIRIE THÉODORE MORGAND, 5, RUE BONAPARTE, A PARIS

1869

大熊猫“身份证” 引自孙前著《大熊猫文化笔记》，五洲传播出版社，2009 年

The giant panda's "ID card". Cited from Sun Qian, *Notes of the Giant Panda Culture*, China Intercontinental Press, 2009

2009 年，将大熊猫“身份证”送到邓池沟天主教堂的联合国教科文组织专家柯高浩（右二）、柯文（右三）、鸟类学家何芬奇（右一）、法国驻成都领事馆领事助理孙雅俊（左三）、四川省旅游局巡视员孙前（左二）及大熊猫文化研究者谭楷（左一）在教堂门前留影 谭楷 供图

UNESCO official Ke Gaohao (2nd from right), Ke Wen (3rd from right), bird scientist He Fenqi (1st from right), French consular assistant Sun Yajun (3rd from left), and inspector with Tourism Administration of Sichuan Province Sun Qian (2nd from left), giant panda culture scholar Tan Kai (1st from left)—they have brought the giant panda's "ID card" to the Dengchigou Catholic Church. This picture was taken in front of the church in 2009. Photo courtesy of Tan Kai

大熊猫“身份证”

1869 年刊载于《巴黎自然历史博物馆之新档案》关于戴维对大熊猫的首次描述和原始记录，被形象地称为大熊猫的“身份证”。2009 年 2 月，有关大熊猫原始文档的复制件被带回中国。大熊猫的“身份证”终于回到了它的故乡。

"ID Card" of the Giant Panda

David's first description and record of the giant panda, published in *New Archives of National Museum of Natural History in Paris* in 1869, is vividly referred to as the "ID card" for the giant panda. In February 2009, copies of those original documents were brought back to China. The "ID card" of the giant panda has finally returned to its homeland.

第二篇
生活

剪纸《家在中国 根在四川》，110cm×70cm（2017 年） 胡宏玲 作品
Paper-cutting: *Home in China,Root in Sichuan*, 110cm×70cm (2017). By Hu Hongling

Part 2
Life

这迷人的造物，是大自然的礼物

珍贵的大熊猫是地球上最孤独的物种之一。

它们数量稀少，总数只有约 2000 只 [1]；

性格孤僻，大多时候都独来独往。

它们藏身高山密林，被称为寄情山水的“竹林隐士”。

[1] 根据全国第四次大熊猫调查结果，截至 2013 年底，全国有野生大熊猫 1864 只，其余为圈养大熊猫。

This cute species is one of the rarest on earth

The giant panda is one of the rarest species on earth.

Its population is small, about 2,000 in total.[1]

It tends to live a solitary and unsociable life in the bamboo groves deep in the mountains,

hence the title of "hermit in the bamboo grove."

[1] According to China's Fourth Giant Panda Census, there were 1,864 pandas in the wild in China. The rest were in captivity.

体操王子（2008 年） 周孟棋 摄

Gymnast (2008). Photo by Zhou Mengqi

第一章 外形

大熊猫为黑白两色，这种搭配在森林里十分显眼，但在雪地里就成了最佳的伪装。它的头部呈圆形，整个面颊也显得十分圆润，它的眼睛很小，瞳孔似猫，因眼睛周边有黑色色块又好像戴上了大大的墨镜。大熊猫的外形，加上肥肥胖胖的身躯，不仅在动物界独树一帜，还极具美学价值。

这一切，都是为了适应自然界的生存规律。稠密的被毛使大熊猫能很好地保持体温，毛的油脂能够防止雨水渗透。厚密的毛发仿佛一层柔软的盔甲，保护着大熊猫免受寒冷、阴湿气候的侵袭。

大熊猫的四肢中前肢稍长、有力，以利攀爬和抓扳竹子。虽然看上去圆滚滚的，十分笨拙，但其全身关节十分灵活，不仅可以用嘴舔咬胯部和尾部，还可以在稠密的竹丛中行走自如，高兴时翻筋斗也十分灵巧。

大熊猫有惊人的咀嚼能力，它的犬齿不太发达，但是臼齿发达，便于磨碎竹叶。粗而短的脖子能够给沉重的头部提供强有力的支撑，浑圆的耳朵、细小的眼睛，这些都是大熊猫适应环境的最佳体现。大熊猫的尾巴比较短，长度约 10 ～ 15 厘米，毛茸茸的，通常贴着身体。

正在爬树的两只大熊猫（2013 年） 谭楷 供图

Two giant pandas are climbing a tree (2013).
Photo courtesy of Tan Kai

Chapter 1 Appearance

工笔画《上树》，90cm×60cm（2013 年）邓维彬 作品

Gongbi painting: *Tree climbing*, 90cm×60cm (2013). By Deng Weibin

The black-and-white appearance of the giant panda is quite an eye-catcher in the forest, but makes for the best camouflage in the snow. Both its head and face are round; its eyes small, like those of the cat; its black circles around the eyes are like a pair of sunglasses; and it has a chubby body. All these make it not only unique in the animal world, but also beautiful to the human eye.

Following the principle of "survival of the fittest," the giant panda wears a dense fur for warmth and grows fat under the fur to prevent rain from penetrating. In other words, its thick fur is like a layer of soft armor, protecting itself from cold and dankness in the mountains.

With its forelegs slightly longer and stronger, this creature is good at climbing and bending bamboo. Though round, fat, and even seemingly clumsy, it is so flexible as to bend to lick its own hindquarters; it ambles quietly through dense bamboo groves, and rolls on the ground when in a good mood.

With less developed canines but well-developed molars, the giant panda has an amazing chewing capacity to grind bamboo leaves. Its thick and short neck provides a strong support for its heavy head. Its ears are round and eyes are small, indicating its perfect adaptation to the environment. Its fluffy tail is short, about 10-15cm long, and usually close to the body.

头部

大熊猫头部看起来圆乎乎的，但仔细观察会发现它的头部器官都有很强的功能性。比如鼻子很短但是鼻腔很大，这是为适应稀薄的高山空气而进化出的一种特征。再比如眼睛的视网膜对光线变化的反应十分敏感，令大熊猫夜行更加灵敏。

Head

The head of a giant panda looks rotund; the features of its sense organs reveal that it has adapted to environmental change. For example, its short nose and large nasal cavities enable it to survive in the thin air in high mountains; and its retina is sensitive to changes in light, enabling it to move about more easily at night.

水彩画《让世界充满爱》

36cm×26cm（2019 年）

德阳市特殊教育学校

欧金花 作品 指导老师 邬殊丽

Watercolor painting:*Fill the World with Love*,36cm×26cm (2019). By Ou Jinhua, from Special Education School of Deyang City, under the instruction of Wu Shuli

国画《大熊猫》

45.5cm×69cm（2019 年）

成都市第十二幼儿园 吴韦彤 作品

指导老师 朱燕

Chinese painting: *The Giant Panda*, 45.5cm×69cm (2019). By Wu Weitong, from Chengdu No.12 Kindergarten, under the instruction of Zhu Yan

第六指

大熊猫有“第六指”，是伪拇指。第六指与其他指配合，能自如地握器掰物。行进时，大熊猫走“内八字”，足向内撇使得它们在竹丛中散步、采食更为方便。大熊猫趾端有坚硬的爪，所以能够在地形陡峭的山间攀爬，行动灵活。

Pseudo-thumb

The giant panda has a pseudo-thumb, or “the sixth toe” to hold and grip objects in cooperation with the other ones. These pigeon-toes enable it to fetch food conveniently, away from the bamboo groves, while the hard claws at the toe tips enable it to adeptly clamber up steep mountains.

中国大熊猫保护研究中心都江堰基地的大熊猫“青青”，绰号“蔓越煤”“青城山小霸王”等，是一个超级网红熊，具有标志性的点绛唇（2018 年） 刘茜 摄

颜色和毛发

大熊猫面部为乳白色，眼周有椭圆形黑色眼圈，耳朵毛色为黑色。鼻尖光滑无毛，呈黑色，裸露在外。腰背与面部同色，为乳白色。前肢至肩以下、后肢的股外侧为黑色。前后足各 5 个脚趾，爪子呈黄白色。足底毛较粗，为黑褐色。大熊猫全身的毛发长度不一，其背部和臀部的毛长约 3.5 ～ 4.5 厘米；肩颈部的毛长约 5 ～ 7 厘米；胸部和腹部的毛最长，约 10 厘米。

Color and Fur

The giant panda has a milky-white face with oval black patches around its eyes. The hair on its ears is also black. The tip of its nose is black, slick and hairless. Its back has the same color as its face, but its forelimbs, shoulder and hind limbs are black. Each of its four paws has five toes and is yellowish-white. The black-brown hair on its planta is thicker. Its hair differs in length, with that on the back and hips about 3.5-4.5cm, that on the shoulder and neck about 5-7cm, and that on the chest and abdomen being the longest at about 10cm.

体重 80 ～ 120 千克
Weight: 80-120 kg

The giant panda Qing Qing, nicknamed Little Briquette, or Little Bully in Mt. Qingcheng, is a super web celebrity. Its red lips are its logo (2018). Photo by Liu Qian

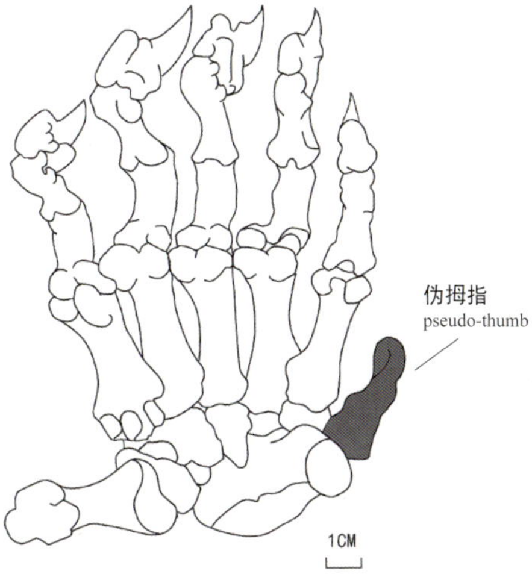

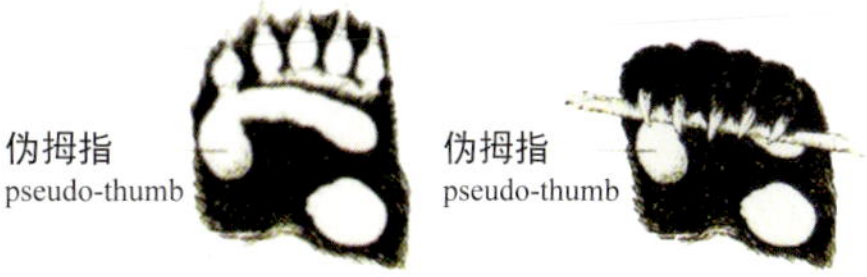

上 大熊猫前掌上的伪拇指 引自魏辅文著《野生大熊猫科学探秘》，科学出版社，2018 年

下 大熊猫前掌上的伪拇指 引自四川省林业厅编《图说大熊猫》，天地出版社，2015 年

(Upper) The pseudo-thumb on the front palm of the giant panda. Cited from Wei Fuwen, *Scientific Exploration of Wild Giant Pandas*, Science Press, 2018

(Below) The pseudo-thumb on its front palm. Cited from *The Giant Panda Pictures* compiled by Forestry Department of Sichuan Province and published by Tiandi Press, 2015

2009 年，棕白色大熊猫“七仔”被发现 （2015 年） 新华社 供图

Qi Zai, one brown-and-white giant panda found in 2009 (2015). Photo courtesy of Xinhua News Agency

1985 年 3 月，在陕西省汉中地区佛坪县岳坝乡大古坪村首次发现了棕白色大熊猫“丹丹” 王磊 摄

Dan Dan, the first brown-and-white giant panda found in Daguping Village, Yueba Township, Foping County, Hanzhong Prefecture, Shaanxi Province, in March 1985. Photo by Wang Lei

2019 年 5 月 25 日，四川卧龙国家级自然保护区管理局对外发布全球首张白色大熊猫照片 四川卧龙国家级自然保护区管理局 供图

The first picture of an all-white giant panda made known to the world on May 25, 2019 by Wolong National Nature Reserve Administration. Photo courtesy of Wolong National Nature Reserve Administration

棕白色和白色大熊猫

大熊猫的颜色除了黑白色，还有棕白色、白色。1985 年 3 月 26 日，一只棕白色患病大熊猫在中国陕西省佛坪自然保护区被发现，后被救护到西安动物园，科研工作者为它取名“丹丹”。这是科学记载中的第一只棕白色大熊猫。据陕西省林业厅统计，截至 2018 年 3 月，世界上有科学记载的棕白色大熊猫发现地点均在陕西秦岭山脉核心地区，一共有 9 次发现野生棕白色大熊猫的记录。

2019 年 4 月中旬，四川卧龙国家级自然保护区内海拔 2000 米左右的一台野外红外触发相机摄录下一张白色大熊猫影像。这只大熊猫通体呈白色，眼睛则是红色，年龄大概在 1 ～ 2 岁，这是有史以来被拍下的第一只白色大熊猫。据专家分析，该白色大熊猫应是一只白化个体。

Brown-and-White and White Giant Pandas

Besides black-and-white pandas, there are also the brown-and-white ones. On March 26, 1985, a sickly brown-and-white panda spotted in Foping Nature Reserve, Shaanxi, was rescued and sent to Xi'an Zoo. Scientists named it Dan Dan, the first ever brown-and-white panda recorded in scientific documents. According to statistics from Shaanxi Province Forestry Department, by March 2018, brown-and-white pandas had all been discovered in the core area of Qinling Mountains in Shaanxi. There had been 9 records of thc sighting of such giant pandas.

In mid-April 2019, a field infrared camera in Wolong National Nature Reserve, Sichuan Province, captured a picture of an all-white Giant Panda at an altitude of about 2,000m. The giant panda, about 1-2 years old, has a white body and red eyes. It is the first all-white panda ever found. According to some experts, it must be an albino.

“七仔”和它的朋友（2011 年） 王磊 摄

Qi Zai and his friend (2011). Photo by Wang Lei

第二章 食性

大熊猫现以高山竹类为主食。找到食物后，它们便悠闲地享受美食。

大熊猫进食像一个绅士。开餐之前，有的大熊猫要先用鼻子去闻竹子，然后才会决定是否品尝。大熊猫嗅觉灵敏，它们很容易找出最可口的竹子。选择好食材后，大熊猫会侧着头将竹子从离地 20 ～ 30 厘米处咬断，再用前肢将竹子拖过来，用嘴咬掉短短一节，便开始吃起来。

大熊猫春季喜欢采食新笋，有“撵笋”[1]的习性。夏季爱吃一两年生的竹竿，秋季时喜欢采食竹叶，冬季则爱食老竹笋。大熊猫不仅挑剔竹子是否鲜嫩，还对竹子的分布区域和种类有要求。它们最爱采食海拔在 1800 ～ 3200 米范围内的各种竹类植物，至于海拔过高的区域，由于受环境影响，竹子生长发育受到抑制，大熊猫不喜欢去采食。

现生大熊猫以吃竹子为主，但也被发现偶食其他食物。在唐家河国家级自然保护区，大熊猫有时会到农田摄食玉米、南瓜、四季豆等，也吃一些野果以及当归等草本植物。大熊猫还会偶食一些动物的尸体，如在四川平武县，红外相机监测发现大熊猫捡食扭角羚的尸体。在宝兴县五龙乡东升村还曾有大熊猫每年都到一农户家“串门认亲”，吃猪骨头。

大熊猫也喜欢喝水，但对于水源水质要求很高。它们的巢域周边一般都会有清泉流水。即使到了天寒地冻的冬天，大熊猫也不会选择吃冰雪解渴，而是每天坚持到流动的溪水处或到还未结冰的泉溪处喝水。

喝水的大熊猫（2010 年）
向定乾 摄

[1] “撵笋”指分布于不同海拔高度的竹子通常从 3 月开始由下而上生长，野生大熊猫追寻新鲜竹笋而发生春季迁徙的行为。

Chapter 2 Dietary Habit

Drinking water (2010).
Photo by Xiang Dingqian

The giant panda feeds on the alpine bamboo. When it happens to get some food, it will enjoy it unhurriedly like a gentleman.

Some pandas will sniff at the bamboo twigs before deciding whether to eat them. With an acute sense of smell, they can find the tastiest bamboo easily. When each finds its food, it bites off the stem 20-30cm above the ground, its head tilted. Dragging the stem over with its forelimbs, it then bites off a short section with its mouth and starts eating the rest.

It is the habit with the giant panda to make seasonal migrations "after fresh bamboo shoots,"[1] fond of new bamboo shoots in spring, one- or two-year-old bamboo stems in summer, bamboo leaves in autumn, and old bamboo shoots in winter. It prefers a variety of bamboo plants growing at the altitude of 1,800-3,200m, and seldom goes to forage in areas of higher-altitude where there are few bamboo groves.

Though it mainly feeds on bamboo, it sometimes eats other food. In Tangjiahe National Nature Reserve, the giant panda sometimes goes to the farmland for corns, pumpkins, green beans and so on. It also eats wild fruits as well as herbs like angelica. It even consumes some dead animal occasionally. For example, in Pingwu County, Sichuan Province, an infrared camera monitor captured the scene of a giant panda eating a dead takin. It is reported to have visited a farmhouse for swine bones in Dongsheng Village, Wulong Town, Baoxing County.

The giant panda is also rather picky about its drinking water. Its home is often built near a running spring or stream. Even in freezing winter, it will get to an unfrozen spring every day, rather than settling for ice and snow.

[1] Bamboos are distributed at different altitudes. Those at the lower altitude start to grow shoots in March and those at the higher altitude do so later. Wild giant pandas migrate in spring to forage fresh shoots.

上 大熊猫在成都大熊猫繁育研究基地的作息时间图（2019 年）

考拉看看 制图 周孟棋 供图 资料来源：成都大熊猫繁育研究基地

下 大熊猫季节活动（2019 年）

考拉看看 制图 资料来源：四川省地方志编纂委员会编《四川省志·大熊猫志》，方志出版社，2018 年

(Upper) The daily schedule of giant pandas in Chengdu Research Base of Giant Panda Breeding (2019). By Koalacan. Data courtesy of Zhou Mengqi. Source: Chengdu Research Base of Giant Panda Breeding

(Below) Seasonal activities of giant pandas (2019). By Koalacan. Cited from *Sichuan Province Annals: Giant Panda* compiled by Sichuan Local Records Editorial Committee, and published by Publishing House of Local Records, 2018

消化系统

大熊猫的消化系统仍保留着食肉动物的特征——单室胃、肠道短，肠道长度约为体长的 4.5 倍。而一般草食动物则为多室胃，肠道的长度是体长的 20 ～ 27 倍。这一差异导致大熊猫吃竹子不能像食草动物（如牛）一样反刍，消化能力较弱。

Digestive System

Its digestive system retains the characteristics of a carnivore: a single-chamber stomach, short intestines about 4.5 times the length of its body. An herbivore usually has a multi-chamber stomach with its intestines about 20-27 times the length of its body. Therefore, though living off bamboo, the giant panda, with a weak digestive capacity, does not regurgitate like the herbivore (the cow, for example).

大熊猫的消化系统

引自四川省档案局、四川省档案学校编《档案见证·大熊猫》，四川人民出版社，2018 年

The digestive system of the giant panda. Cited from *Archives Witness: The Giant Panda* compiled by Sichuan Province Archives Bureau and Sichuan Province Archives College, and published by Sichuan People's Publishing House, 2018

季节习性

大熊猫活动的坡向与季节关系较大，冬春季大熊猫喜欢在阳坡面活动，阳坡面日照充足，积雪易融，方便其晒太阳；夏秋季则喜欢在阴坡面活动，阴坡面一般水源丰富，竹笋茁壮，大熊猫在避暑的同时还能享受美食。

Seasonal Habits

A slope is where the giant panda frequents. In winter and spring, it prefers a south-facing slope with enough sunshine, which helps to melt the snow. In summer and autumn, it loves a north-facing slope where abundant water and exuberant bamboo are available.

大熊猫采食竹竿 引自魏辅文著《野生大熊猫科学探秘》，科学出版社，2018 年

A giant panda is eating bamboo sticks. Cited from Wei Fuwen, *Scientific Exploration of Wild Giant Pandas*, Science Press, 2018

食性变化

大熊猫何时开始食竹？学界有多种看法。从始熊猫牙齿构造上看，应为杂食性动物。它的生活环境与现生的大熊猫大不相同，竹子可能不是始熊猫的主要食物来源。小种大熊猫在齿型构造上与现生大熊猫相似，因此，大熊猫家族以竹为主食，可能始于小种大熊猫。然而，一项基于 12 只古代大熊猫骨胶原蛋白的研究结果表明，古代大熊猫可能具有更多不同的食物来源，竹子并不是它们唯一的食物，大熊猫以竹为主食的生活方式可能只有几千年的历史。

Dietary Changes

When did the giant panda begin to live on bamboo? Opinions vary in academia. The teeth structure of *Ailuaractos* (primal panda) indicates that they were probably omnivorous. Their living environment was quite different from that of the modern giant panda, which means bamboo was probably not its staple food. The teeth of *Ailuropoda microta* (pygmy giant panda) resemble those of the present-day giant panda in structure, therefore, the origin for the giant panda selecting bamboo as its staple food traces back to *Ailuropoda microta*. However, findings from a study based on bone collagen of twelve ancient giant pandas suggest that the ancient panda might have relied on more diverse food sources instead of bamboo as the sole food. So, it seems only a few thousand years have passed since the giant panda began to depend chiefly on bamboo.

国画《秋韵》，137cm×69cm（2018 年） 刁华甫 作品

Chinese painting: *Autumn Rhythm*, 137cm×69cm (2018). By Diao Huafu

竹林隐士

大熊猫畏惧酷热，一般情况下，习惯在湿度 70% ～ 80%，海拔 1800 ～ 3200 米的深山竹林中生活，由此得到一个“竹林隐士”的雅号。

除了深山竹林，沟尾平塘、山腰平塘和窝凼等处也是大熊猫喜爱的活动场所。大熊猫还喜欢在海拔较高、僻静的河谷地带游玩，它们也常独自游荡于斜坡、缓坡、小石包和小山脊等处。

Hermit in the Bamboo Grove

Afraid of extreme heat, and fond of an environment of 70%-80% humidity, the giant panda usually dwells in the bamboo groves deep in the mountains at 1,800-3,200m in altitude, hence the nickname “hermit in the bamboo grove.”

While inhabiting the bamboo grove in the deep mountains, the giant panda also roams slopes, mounds and ridges, ranges from high-altitude to secluded valleys, and haunts lakes, pools and ponds in valleys or hillsides.

大熊猫爬树

大熊猫看似笨拙，但行动十分灵活，无论是觅食还是爬树，都是高手。如果受到惊吓，它们还会急速小跑。大熊猫前肢比后肢发达，在爬树时前肢贴在树身，慢慢向上，直至后肢直立，然后爬上大树。在下树时头仍向上慢慢下退，直到接近地面时，才敏捷地纵身跃下。

Tree Climber

Clumsy though it appears, a giant panda is in truth very agile, whether it is foraging or climbing a tree. If startled, it will trot, its forelimbs better developed than its hind limbs. When climbing a tree, it rises slowly , forelimbs gripping the trunk, until its hind limbs are off the ground, and then scrambles up to the desired position in the tree. When descending, it keeps its head upwards and moves downwards slowly until near the ground, where it jumps off.

左 各有高招（2007 年） 周孟棋 摄
右 绢本《天之骄子》
127cm×212cm（2018 年） 王申勇 作品
(Left) Climbing (2017). Photo by Zhou Mengqi
(Right) Chinese silk scroll painting: *God's Favored Ones*, 127cm×212cm (2018). By Wang Shenyong

草丛中那惊鸿一瞥，就是爱情（2000 年） 周孟棋 摄

Courting (2000). Photo by Zhou Mengqi

1. 粉彩笔画《熊猫恰恰》，39cm×27cm（2019 年） 四川省直属机关实验婴儿园 王智晗 作品 指导老师 郑江
2. 油性粉彩画《吃在成都——火锅》，54cm×38cm（2019 年） 四川省直属机关实验婴儿园 周斯年 作品 指导老师 郑江
3. 油性粉彩画《飞翔太空》，54cm×38cm（2019 年） 四川省直属机关实验婴儿园 袁骁洲 作品 指导老师 郑江
4. 水彩画《快乐信使》，54cm×38cm（2019 年） 四川省直属机关实验婴儿园 曾雯聆 作品 指导老师 郑江
5. 线描画《和谐号》，39cm×27cm（2019 年） 四川省直属机关实验婴儿园 蒲世悦 作品 指导老师 于小莉

第三章　情愫

每年 3—5 月是大熊猫交配的季节。在渐浓的春光下，原本习惯独居的野生大熊猫开始通过气味和叫声相互联系。

雄性大熊猫在发情期的反应比雌性强烈，它们从 12 月开始，将尿液撒在稍高的植物或岩石上作为气味标志，通过气味向周围的雌性大熊猫发出信号，到了 4 月，这种行为越发频繁。大熊猫还通过叫声相互联系。一只雌性大熊猫发情能够引来数只雄性大熊猫，这个时候，雄性大熊猫不得不通过“决斗”来争夺配偶。

野生大熊猫交配持续时间通常在 1 ～ 3 分钟，一次交配活动的结束多以雌性大熊猫反身攻击雄性大熊猫并逃离为标志。交配完成后，雌性大熊猫会与雄性大熊猫分开，独栖生活。

受环境、气候、营养水平、健康状况、饲养管理等众多因素影响，圈养大熊猫发情很难。20 世纪 90 年代之前，由于对大熊猫的自然习性认识不足，圈养大熊猫多数不能表现出正常的发情行为。圈养雄性大熊猫的自然交配率还不到 10%，雌性约为 50%。

20 世纪 90 年代之后，这一问题得到解决。

1. Pastel painting: *Dancing*, 39cm×27cm (2019). By Wang Zhihan from Sichuan Province Experimental Kindergarten, under the instruction of Zheng Jiang

2. Oil pastel painting: *Chengdu Hot Pot*, 54cm×38cm (2019). By Zhou Sinian from Sichuan Province Experimental Kindergarten, under the instruction of Zheng Jiang

3. Oil pastel painting: *Travel in Space*, 54cm×38cm (2019). By Yuan Xiaozhou from Sichuan Province Experimental Kindergarten, under the instruction of Zheng Jiang

4. Watercolor painting: *The Messenger*, 54cm×38cm (2019). By Zeng Wenling from Sichuan Province Experimental Kindergarten, under the instruction of Zheng Jiang

5. Sketching: *CRH*, 39cm×27cm (2019). By Pu Shiyue from Sichuan Province Experimental Kindergarten, under the instruction of Yu Xiaoli

Chapter 3 Mating

The mating season for the giant panda is between March and May. With the advent of bright spring, the wild giant panda, typically detached from its peers, begins to attract the attention of the other sex by leaving smells and giving cries.

The male is more proactive when in heat than a female. From December onward, it starts to leave urine on a tall plant or rock as a signal for any female around. This behavior becomes more frequent in April. A female in heat can attract several males. In that case, the males have to “duel” for the chance of mating.

The wild giant panda’s mating usually lasts one to three minutes. Once the copulation is finished, the female will attack the male and flee, each going its own way.

Environment, nutrition, health condition, breeding management and so on have a direct influence on the giant panda’s oestrous behavior; it is difficult for a sexually mature one in captivity to be in heat. Before the 1990s, most of those in captivity could not go into oestrus. Less than 10% of males mate naturally, while about 50% of females mate naturally.

The problem was not solved until afterward.

拥吻（2012 年） 周孟棋 摄

Smooch (2012). Photo by Zhou Mengqi

纸本水墨丙烯《熊猫家族》

97cm×60cm（2019 年） 米金铭 作品

Chinese painting of ink and acrylic: *A Giant Panda Family*, 97cm×60cm (2019). By Mi Jinming

性成熟

圈养大熊猫的性成熟年龄要比野生大熊猫早 1 ～ 2 年。人工饲养条件下，大熊猫个体不同，其性成熟年龄也存在差异。雄性大熊猫一般在 5.5 ～ 6.5 岁性成熟，而最晚完成交配的个体年龄为 8.5 岁，最早完成交配的个体年龄为 4.5 岁；雌性大熊猫一般在 4.5 ～ 5.5 岁性成熟，而最早的 2.5 岁就有了比较典型的发情行为，最晚的在 7.5 岁左右才出现这种行为。雄性大熊猫一般比雌性大熊猫晚 1 年左右性成熟。

Sexual Maturity

The giant panda in captivity becomes sexually mature one to two years earlier than the wild one. A captive male normally reaches sexual maturity at the age of 5.5-6.5, though some estrous at 4.5 while others at 8.5. The female turns sexually mature at the age of 5.5-6.5, though some are in heat at 2.5 while others at about 7.5. The time for a male to get sexually mature is generally about one year later than that of a female.

大熊猫用不同姿势将尿液或肛周腺分泌物遗留在活动范围内，气味会携带个体大小、性别、繁殖情况等信息 引自赵学敏编《大熊猫——人类共有的自然遗产》，中国林业出版社，2006 年

Giant pandas leave urine and perianal gland secretions in different positions in their ranges. Their odor carries information of body size, sex, breeding status, etc. Cited from Zhao Xuemin, *The Giant Panda: Natural Heritage of Humanity*, China Forestry Publishing House, 2006

发情期

人工圈养的 10 岁以下的雌性大熊猫发情期较长，一般能够持续 2 ～ 3 周，10 ～ 18 岁的雌性大熊猫发情期一般能够持续 1 ～ 2 周，19 岁以上的雌性大熊猫发情期比较短，通常是 5 ～ 7 天。

Estrous Period

In captivity, a young female under the age of 10 has a longer period of heat, generally lasting 2-3 weeks, a middle-aged female between the age of 10 to 18 has 1-2 weeks, and an older female over the age of 19 has only 5-7 days.

大熊猫妈妈叼着大熊猫宝宝 引自谭楷著《熊猫的故事》，五洲传播出版社，2015 年

A cub in its mother's mouth. Cited from Tan Kai, *Story of the Giant Panda*, China Intercontinental Press, 2015

1

2

1. 剪纸《武侯祠熊猫》，直径 20cm（2005 年） 龙玲 作品
2. 剪纸《花瓶熊猫》，22cm×17cm（2015 年） 林琼华 作品
3. 剪纸《爱》，60cm×42cm（2019 年） 何小锵 作品
4. 剪纸《熊猫萌萌哒》，27cm×21cm（2016 年） 游琴舒 作品
5. 剪纸《福寿双全》，20cm×22cm（2016 年） 高慧兰 作品

第四章　育幼

雌性大熊猫在怀孕后，有约 5 个月的妊娠期[1]。在这期间，需通过大量进食来满足自己和腹中宝宝的营养需求。到了临产前几天，则基本不吃不喝。

妊娠后期，雌性大熊猫活动逐渐减少且嗜睡，但还必须为自己营造一间温馨的“产房”。野生大熊猫产崽多选择在大树空洞，也有在石洞或大树干基部筑巢的。为了让巢穴温暖舒适，大熊猫还找来树枝作为铺垫。“产房”内部往往十分黑暗，但温度稳定。哺幼期间，雌性大熊猫常常用背部堵住洞口，这样做不仅可避风寒，还可使洞内温度升高。只要洞口被遮住，洞内便是绝佳的产育场所。

完成这些准备，雌性大熊猫的活动减少，开始在产房内静静等待宝宝出生。分娩前，生理反应会让大熊猫烦躁不安，因腹部用力和呻吟，嘴里会发出“嗯、嗯”的叫声，有的雌性大熊猫还发出“咩咩”的叫声。雌性大熊猫一般是 7—9 月分娩。

雌性大熊猫分娩时长一般为 3 ～ 5 个小时。顺利产下幼崽后，疲惫的雌性大熊猫在 2 ～ 3 天内不吃不喝。在哺乳时，它用前肢抱着幼崽坐着，幼崽趴在自己的腹部吸奶，吸完一边，它用前肢托着幼崽，并用嘴协助把幼崽拉到另一边再吸。刚出生的大熊猫幼崽叫声特别大，在一个多月里，雌性大熊猫要一直把它抱在怀里。大熊猫宝宝啼叫时，雌性大熊猫会不断舔它的身体，有时甚至会衔起走动。

[1] 由于大熊猫胚胎有延迟着床的生理特点，妊娠期变化范围较大，从 85 ～ 185 天不等，极个别大熊猫妊娠期长达 324 天。

3

4

5

1.Paper-cutting: *The Giant Panda in the Temple of Marquis*, 20cm in diameter (2005). By Long Ling

2.Paper-cutting: *The Giant Panda on the Vase*, 22cm×17cm (2015). By Lin Qionghua

3.Paper-cutting: *Love*, 60cm×42cm (2019). By He Xiaoqiang

4.Paper-cutting: *Cute Giant Panda*, 27cm×21cm (2016). By You Qinshu

5.Paper-cutting: *Felicity and Longevity*, 20cm×22cm (2016). By Gao Huilan

Chapter 4 Breeding

The female giant panda has a five-month gestation period[1],during which the mother to-be needs a lot of food to provide enough nutrition for the baby and itself. A few days before delivery, it barely eats or drinks.

At a later stage of gestation, it becomes less active, and even lethargic. But before that occurs, it needs to create a warmer den for delivery. A wild female prefers to do it in a hollow tree, in a rock cave or at the base of a tree trunk. The would-be mother uses twigs and branches as bedding to keep the den warm and comfortable. It is usually very dark inside, but the temperature is stable. Before the baby is born, the mother often blocks the entrance to the delivery chamber with her back, which helps to keep out the cold wind. With the entrance covered, the cave is an ideal place to give birth in.

Once completing the preparations above, the pregnant panda slackens itself, awaiting the arrival of its baby. But before delivery, it tends to be irritated and uneasy. Groaning and straining from the abdomen, it sometimes bleats like a sheep. Giant pandas usually give birth between July and September.

It usually takes the female 3-5 hours to give birth. After that, the exhausted mother neither eats nor drinks for 2-3 days. When nursing its baby, the mother sits and holds it with its forelegs while the baby lies on its stomach suckling its breasts. Once the cub finishes suckling on one side, the mother will change it to the other side with its mouth. The newborn cries loudly and the mother has to hold it in her arms from time to time, with this lasting more than one month. When the baby cries, the mother will lick its body and sometimes even pick it up in its mouth to walk around.

[1]Owing to the delayed implantation of giant panda embryos, the gestation period ranges from 85 to 185 days, with the longest one up to 324 days.

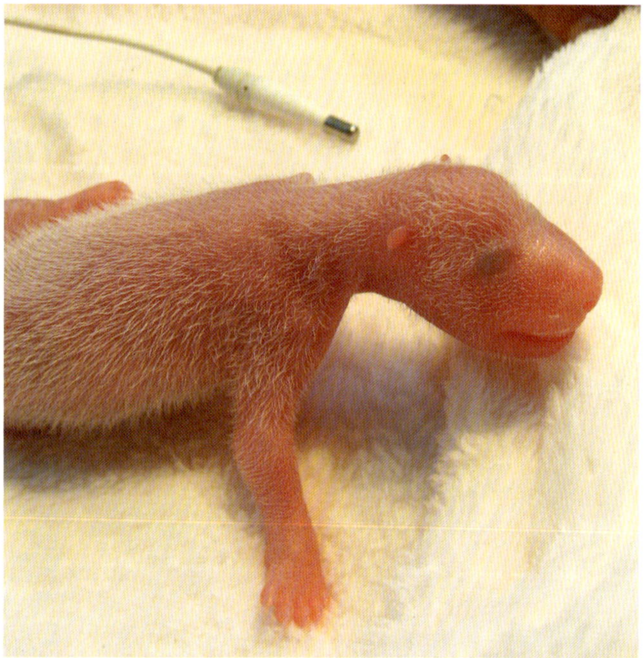

刚出生的大熊猫宝宝
引自谭楷著《熊猫的故事》，五洲传播出版社，2015 年

A Newborn Cub. Cited from Tan Kai, *Story of the Giant Panda*, China Intercontinental Press, 2015

1988 年秋季，宝兴县蜂桶寨自然保护区管护员正在精心喂养大熊猫幼崽 高华康 摄

A keeper of Fengtongzhai National Nature Reserve was feeding a cub in Autumn 1988. Photo by Gao Huakang

初生大熊猫

20 世纪 80—90 年代，圈养大熊猫幼崽的成活率仅 30% 左右。20 世纪 90 年代后，通过科研人员不断努力，大熊猫幼崽的成活率提高到 90% 以上。

从圈养大熊猫来看，刚出生的大熊猫体重约 140 克，双眼紧紧地闭着，体表呈肉红色，长着白色的短胎毛。它生长速度很快，1 个月体重就能长到初生时的 7 倍，半年能达 11 千克。

Newborn Giant Panda

In the 1980s to the 1990s, the survival rate of cubs born in captivity was only about 30%. But since 1990, it has increased to more than 90%, thanks to the persistent efforts of researchers.

With a weight of only about 140g, the captive baby is very fragile. Its eyes are closed; its pale red body is covered with white and short lanugos; it grows so fast that in a month it will be 7 times weightier than at birth, and in six months will reach 11kg.

1982 年 9 月 30 日，四川省宝兴县硗碛藏族乡的一只野生大熊猫产下一只幼崽 高华康 摄

On September 30, 1982, a wild giant panda birthed a cub in Qiaoqi Tibetan Township, Baoxing County, Sichuan Province. Photo by Gao Huakang

母兽产后活动

产崽后，雌性大熊猫会安静地待在自己为育幼所挑选的巢穴中。在野外，产后 15 天左右雌性大熊猫已可短暂外出，25 ～ 30 天后可较长时间离开幼崽外出活动。

Postpartum Activities

After delivery, the mother stays quiet in the den. In the wild, it is not until 15 days later that it leaves the den and remains away for a short while, and not until 25-30 days later that it leaves its cub alone in the den for a longer period of time.

大熊猫的生长过程
Developmental phases of the giant panda

大熊猫的生长过程（2019 年） 考拉看看 制图 四川省档案馆、四川省林业和草原局，周孟棋 供图

Developmental phases of the giant panda (2019). By Koalacan. Photo courtesy of Sichuan Province Archives, Sichuan Province Forestry and Grassland Administration and Zhou Mengqi

45天
One and half months old

1岁
One year old

成长过程

大熊猫在出生后 40 天左右（有的约 2 个月），才逐渐睁开双眼看到这个新奇的世界；80 ～ 100 天时，它开始长牙，并能用四肢支撑身体，勉强站立；约 120 天时便可走动；到了 130 ～ 150 天时，它变成了喜欢打滚、玩耍的“熊孩子”。从出生到 1.5 岁，这是大熊猫的幼年期；1.5 岁之后，大熊猫便开始独立生活；从 1.5 岁到性成熟（约 5 岁），这是大熊猫的亚成年期；5.5 ～ 6.5 岁（或雌性 6.5 岁、雄性 7.5 岁）步入成年期，开始寻找配偶，繁育后代。

Growth

It is about 40 days (or two months in some cases) before the newborn panda can open its eyes; 80-100 days before it begins to grow teeth and manages to stand on its own feet; about 120 days before it can toddle around; 130-150 days before it becomes a “naughty kid” enjoying itself by rolling and doing other forms for fun. The infancy period is from birth to the age of 1.5, followed by independence from its mother. The period from the age of 1.5 to when it turns sexually mature (about 5 years old), is the sub-adult period. It becomes an adult at the age of 5.5-6.5 (or the female at 6.5 and the male at 7.5), beginning to look for a mate for reproduction.

第三篇

环境

盆景《戏竹》，120cm×70cm（2019 年） 陈志贵 作品

Bonsai: *Having Fun in the Bamboo Grove*, 120cm×70cm (2019). By Chen Zhigui

Part 3
Environment

这是应许之地，是它的乐园

随着地壳运动和气候变化，
中国西部山区成为珍稀动植物的“诺亚方舟”。
在这片广袤的森林里，
大熊猫获得了充足的食物、纯净的水源和安全的栖息地。
它们如同“黑白精灵”与其他动物和谐共生，
共同构建了生机盎然的生态系统。

This is a promised land, a paradise for the giant panda

Tectonic movement and climate change have turned the mountainous regions
in western China into a Noah's Ark for rare animals and plants.
In the vast forests,
the giant panda can find sufficient food and clear water as well as find its safe den.
Such a black-and-white spirit of nature shares its habitat with other animals,
thus forming a dynamic and vibrant ecosystem.

第一章 分布

如果从空中俯瞰现生大熊猫的分布区域，它呈弧形。它的北部是秦岭山系，西北部是岷山山系，西部是邛崃山山系，西南段是大、小相岭山系和凉山山系。这些分布区域集中在西秦岭褶皱带与昆仑褶皱带的交会之地，在自然地理特征上存在着显著差异。

和历史上的大熊猫分布区域相比，如今的大熊猫分布区域已缩小了不少。历史上，大熊猫的分布范围甚广，大体位于北纬 23°～41°，东经 95°～120°之间。这一区域属亚热带和暖温带，气候温暖湿润，森林密布，竹类繁多，适合大熊猫生存。

随着地质结构和气候变化，森林面积减少，大熊猫栖息地面积锐减。现生大熊猫的分布区域纬度仅跨越近 6°，经度仅跨越近 7°。现集中分布在青藏高原东缘、川西北部山地、甘肃南部山地和秦岭山地的部分地区，总面积约 2.58 万平方千米。

这 2.58 万平方千米的区域是大熊猫最后的“诺亚方舟”。这里地形复杂，海拔高差大，形成了多样的气候环境和复杂的生态类型。

在这片广袤的区域里，暖温带山地森林与亚热带山地森林相互交错，为大熊猫种族的延续保存着最后火种。

国画《相伴》，180cm×180cm（2017 年）
刘进全 作品

Chapter 1 Distribution

Chinese painting: *Companions*, 180cm×180cm (2017).

By Liu Jinquan

If you look down at the distribution areas of today's giant panda from the sky, a curved outline comes into view, with the Mountain Ranges of Qinling in the north, Minshan in the northwest, Qionglai in the west, Daxiangling, Xiaoxiangling and Liangshan in the southwest. Such a habitat, with its remarkable physical and geographical features, concentrates upon the intersection of West Qinling and Kunlun fold belts.

Compared with that of the past, the habitat for today has been largely reduced. Throughout history, the giant panda has been generally distributed between 23°-41°N and 95°-120°E, an area in the north subtropical and warm temperate zones. With a warm and humid climate, thick forests and varieties of bamboo, this region makes a perfect giant panda habitat.

As the geological structure and climate evolve, the forested area has narrowed, and so has the habitat for the giant panda. Within a latitudinal and longitudinal extent of 6-7 degrees, today's giant panda mainly lives at the edge of the eastern Qinghai-Tibet Plateau and in the mountains in northwestern Sichuan, south Gansu and part of the Qinling Mountains.

This area of approximately 25,800 km² stands as the "Noah's Ark". Owing to the complex terrain and significant height differences above sea level, there exist varied climates and diverse ecotypes.

In this vast expanse of land, warm temperate montane forests overlap with subtropical ones, which provide the last hope of reproduction for the giant panda.

地质变迁

在大熊猫漫长的演化过程中，不断隆升的青藏高原成为大熊猫分布范围拓展的一道天然屏障。青藏高原改变了亚洲内部的气候环境和地形地貌，为大熊猫提供了各种可能的生存地。

Tectonic Evolution

During the long evolutionary process of the giant panda, the ever-uplifting Qinghai-Tibet Plateau serves as a natural barrier for its scope of distribution. Qinghai-Tibet Plateau has changed the climate environment and landform within the Asian Continent, providing the giant panda with a variety of possible places for existence.

地势走向

在中国地势的三级阶梯中，现生大熊猫的分布区域位于第二阶梯。这种“阶梯式”的地形，主要是在东西向和北东向两级构造线的影响下形成，反映了地壳差异性的构造运动。这些山体组成了自西北向东南倾斜的空间分布格局，山脉在排列上几乎与地质构造走向一致。

Topography

The area of distribution for today's giant panda stands on the second step of China's three-step "staircase-like" topography. This terrain has formed mainly under the influence of the east-west and north-east tectonic movement which has rendered different configurations on the crust. The mountains incline from northwest to southeast, and are practically concurrent with the geological structure.

白云缭绕在山间（2008 年）周孟棋 摄

Misty mountains (2008). Photo by Zhou Mengqi

中国大熊猫保护研究中心卧龙核桃坪基地（2006 年） 中国大熊猫保护研究中心 供图
Wolong Hetaoping Base of China Conservation and Research Center for the Giant Panda (2006). Photo courtesy of China Conservation and Research Center for the Giant Panda

中国大熊猫保护研究中心卧龙核桃坪基地（2019 年） 中国大熊猫保护研究中心 供图
Wolong Hetaoping Base of China Conservation and Research Center for the Giant Panda (2019). Photo courtesy of China Conservation and Research Center for the Giant Panda

左／右 20 世纪 90 年代，天然林禁伐之前四川省平武县采伐林木的情景 平武县林业局 供图

(Left/Right) Lumbering in Pingwu County in the 1990s, before the ban on logging of natural forests. Photo courtesy of Pingwu County Forestry Administration

人类影响

许多人类活动，比如砍伐、开垦、放牧、基础设施建设等，破坏性地改变着森林生态环境，造成大熊猫栖息地面积减少、整体性被破坏。这导致大熊猫分布范围缩小，并被阻隔为彼此不能相通的不同局域种群，更糟的是，形成了灭绝风险极高的若干孤立小种群。

Human Encroachment

Human activities, such as logging, reclamation, grazing, and infrastructure construction, have destructively changed the forest ecology. Consequently, the habitat for the giant panda has been shrinking, affecting the populations of different localities abilities to communicate with each other, and worse still, landing the small isolated ones on the verge of extinction.

2017 年，在四川省崇州市鞍子河镇境内被红外相机拍摄的大熊猫 鞍子河国家级自然保护区 供图

A giant panda photographed by an infrared camera in Anzihe Township, Chongzhou City, Sichuan Province in 2017. Photo courtesy of Anzihe National Nature Reserve

自然资源

大熊猫分布区内的原始森林资源、竹林资源、水资源以及其他各种自然资源，对于大熊猫来说十分重要。离开了这些资源，大熊猫将无法生存下去。

Natural Resources

It is vital for the giant panda to have various kinds of natural resources in the habitat, including the primeval forests, bamboo groves and water resources. Without all these, it can by no means live on.

第二章　山系

中国西部地区的六大山系是现生大熊猫分布的区域。它包括秦岭山系、岷山山系、邛崃山山系、大相岭山系、小相岭山系和凉山山系。其中，大熊猫种群数量较大的是岷山山系、邛崃山山系和凉山山系。

六大山系横跨亚热带和暖温带两个气候带，气候温暖，雨水丰沛，包含针叶林、阔叶林、针阔叶混交林，更有 40 多种大熊猫可食竹。

对于大熊猫而言，森林植被类型是决定它们生存的重要因素。挺拔繁密的大树为林下竹类健康成长提供了条件，为大熊猫提供了休憩和躲避敌害的重要场所，而中空的树洞又是繁育后代最好的“产房”。

六大山系的地形条件也影响了大熊猫的生活方式。由于海拔高度、气温、水源、植被类型等自然条件与食物分布密切相关，大熊猫会随季节变化在不同海拔地带和植被类型中来回迁移，如同崇山峻岭中的“游牧民族”。

以岷山山系四川境内段为例，在每年 2—5 月，大熊猫的活动范围主要集中在海拔 2200 ～ 2600 米的区域；5—7 月，活动范围的上限往往会扩展到海拔 3400 米左右；到了 8—9 月，活动范围则基本在海拔 2400 ～ 3500 米之间。可见，随着气温逐渐升高，大熊猫的活动范围也随之拓展。

在大熊猫生活的六大山系中，森林资源丰富，主食竹种类繁多，各种珍贵的飞禽走兽生存其间。凭借得天独厚的自然条件，这一片区域成为大熊猫繁衍生息的一方净土。

国画《崛》，244cm×122cm（2008 年）　刘中　作品
Chinese painting: *Rising*, 244cm×122cm (2008). By Liu Zhong

Chapter 2　Mountain Ranges

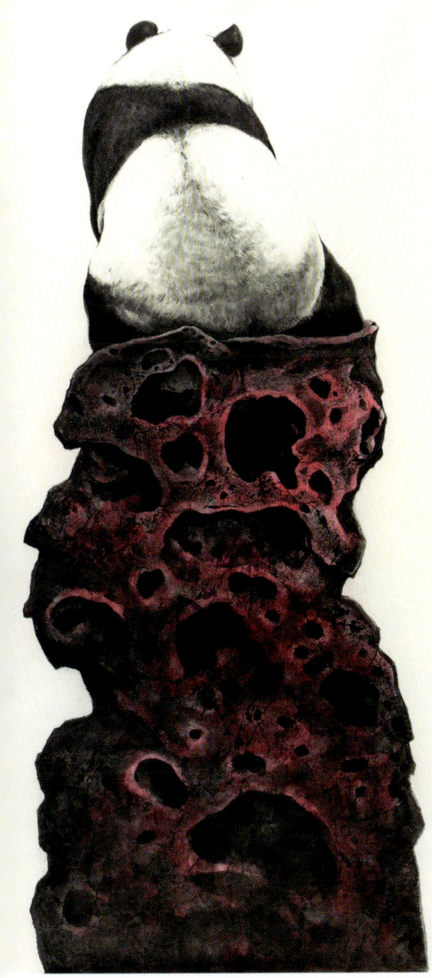

《雍》，244cm×122cm（2009 年）　刘中　作品
Chinese painting: *Harmony*, 244cm×122cm (2009). By Liu Zhong

The six major mountain ranges, located in western China as the main habitations for the giant panda are Qinling, Minshan, Qionglai, Liangshan, Daxiangling and Xiaoxiangling. Among them, Minshan, Qionglai and Liangshan boast the largest populations.

The six major mountain ranges above cover both the subtropical and warm temperate zones, where warm climate and ample rainfall nurture the coniferous, broad-leaved and coniferous-broad-leaved forests and more than 40 kinds of edible bamboos for the giant panda.

Diverse vegetation in forests are the determinants for panda subsistence. Towering and thick trees help bamboo grow healthy and sound in the forest and provide desirable shelters for the giant panda, and protect against enemies as well. And the tree hollows can serve as delivery dens for the females.

The terrain there also has an influence on its lifestyle. As altitude, temperature, rainfall and vegetation types are closely related to its food distribution, the giant panda has to migrate with the seasons.

Take the Minshan Mountains in Sichuan as an example. The activity scope for the giant panda centers in the area of 2,200-2,600m above sea level from February to May; while from May to July, the upper limit of the scope often extends to about 3,400m above sea level. And from August to September, this species stays between 2,400-3,500m above sea level. As the temperature increases, its activity scope also expands.

In these six major mountain ranges for the giant panda, forests are lush, staple edible bamboo is diversified, and all kinds of precious animals and birds coexist-these conditions turn this vast expanse of land into a paradise.

秦岭山系，引自四川省林业厅编《图说大熊猫》，天地出版社，2015 年

Qinling Mountain Range. Cited from *The Giant Panda Pictures* compiled by Forestry Department of Sichuan Province and published by Tiandi Press, 2015

秦岭山系

横亘于渭河和汉江之间，东西绵延近 500 千米，南北宽 100 ～ 150 千米。其主峰太白山海拔高度 3771.2 米，整体地势为东低西高、北陡南缓。大熊猫分布在秦岭南坡中段一带，这里坡度较为平缓，峡谷河流源远流长。秦岭山系四川部分的野生大熊猫仅分布在青川县姚渡镇。

Qinling Mountain Range

It lies between Weihe River and Hanjiang River, stretches nearly 500km from east to west and 100-150km from south to north. Its highest peak, Mount Taibai, stands 3,771.2m above sea level. It descends from east to west, steep in the north and gentle in the south. The giant panda resides in the middle area of the southern slopes, where rivers flow through perennially. Where Qinling borders on Sichuan, the wild giant panda only inhabits the area in Yaodu Township, Qingchuan County.

唐家河国家级自然保护区（2007 年） 董磊 摄

Tangjiahe National Nature Reserve (2007). Photo by Dong Lei

岷山山系

在秦岭的西面，它的最北端在中国甘肃省岷县，最南端在中国四川省都江堰市，西面到郎木寺、毛尔盖，东至宕昌、武都、青川一线，从南到北长达 500 千米，因而有“千里岷山”的说法。岷山山系的主峰是雪宝顶，海拔高度 5588 米。岷山北段、中段和南段是野生大熊猫的主要分布区。

Minshan Mountain Range

Located west of Qinling, it runs 500km (1000 *li*, hence the nickname of “1000 *li* Minshan”) from Dujiangyan City, Sichuan in the south, to Minxian County, Gansu in the north; it reaches Langmu Temple and Maoergai in the west, and extends to Tanchang, Wudu and Qingchuan in the east. Mount Xuebaoding, its highest peak, rises 5,588m above sea level. The giant panda haunts mainly the north, middle and south areas of Minshan.

四姑娘山（2015 年） 何英 摄

Mt. Siguniang (2015). Photo by He Ying

邛崃山山系

南北走向，南北绵延约 250 千米，是岷江和大渡河的分水岭。邛崃山的主峰是四姑娘山，最高海拔 6250 米。邛崃山山系由于海拔较高，自然带垂直分异明显，生物多样性复杂，中段、南段成为野生大熊猫主要分布区域。

Qionglai Mountain Range

It runs in the north-south direction, extending around 250km as the watershed of the Minjiang and Dadu Rivers. Its highest peak, National Scenic Area of Mt. Siguniang, stands at 6,250m above sea level. This high altitude results in biodiversity and distinct natural layering of ecosystems at different elevations, thus making Qionglai the major habitat for the giant panda.

大相岭山系

西起二郎山，东至峨眉山，呈东西走向，绵延 200 余千米，平均海拔 3000 米。坐落其间的大渡河金口大峡谷全长约 26 千米，于 2005 年被评为中国最美十大峡谷之一。野生大熊猫主要分布于大相岭山系的东北坡。

Daxiangling Mountain Range

Running in the west-east direction between Mt. Erlang and Mt. Emei, it stretches over 200km with an average altitude of 3,000m above sea level. The 26-kilometer-long Jinkou Grand Canyon along the Dadu River that rolls through these mountains, was listed in 2005 as one of the ten most beautiful gorges in China. The wild giant panda is mainly distributed on the northeast slope of the Daxiangling.

大相岭自然保护区（2015 年） 周孟棋 摄

Daxiangling National Nature Reserve (2015). Photo by Zhou Mengqi

石棉县栗子坪彝族孟获村月亮湖秋色（2017 年） 刘玉梅 摄

Autumn of Moon Lake in Menghuo Village of Yi Nationality in Liziping, Shimian County (2017). Photo by Liu Yumei

小相岭山系

位于石棉与西昌之间，呈南北走向，绵延超过 100 千米，是大雪山延伸出的余脉。海拔 4500 米的主峰俄尔则俄峰，直耸云霄。野生大熊猫主要分布于小相岭山系东段。

Xiaoxiangling Mountain Range

Located between Shimian and Xichang as an extension of the Daxue Mountains, Xiaoxiangling runs in a north-south direction for over 100km. Its highest peak Erzer rises 4,500m above sea level. The wild giant panda is mainly distributed in its east section.

凉山山系

位于横断山脉东北缘，是四川盆地和云贵高原过渡地带，大部分山体海拔在 2000 ～ 3000 米。古代南方丝绸之路也蜿蜒其中。不论是大熊猫还是人类，都在这里留下了频繁活动的印记。野生大熊猫主要分布于凉山山系中部。

Liangshan Mountain Range

At an altitude of 2,000-3,000m above sea level, it is located at the northeastern edge of Hengduan Mountains and connects the Sichuan Basin and Yunnan-Guizhou Plateau. The Ancient Southern Silk Road, which zigzags through it, still bears the footprints of both human and the giant panda. The wild giant panda inhabits its middle section.

马边大风顶国家级自然保护区 引自四川省林业厅编《图说大熊猫》，天地出版社，2015 年

Mabian Dafengding National Nature Reserve. Cited from *The Giant Panda Pictures* compiled by Forestry Department of Sichuan Province and published by Tiandi Press, 2015

第三章 竹类

在大熊猫分布区，生长着40多种大熊猫可食用竹类，其中主食竹30多种，包括分布最广的箭竹属，主要分布在亚洲东部和非洲的玉山竹属，生长在长江流域与南方山地丘陵地带的箬竹属，主产于喜马拉雅南麓及中国西南部的镰序竹属，分布于中国四川省以及湖北省西部、陕西省南部、甘肃省南部等地的巴山木竹属，分布于中国西南部的筇竹属，分布于中国、日本、缅甸、印度等国家的寒竹属，分布于中国四川盆地西部边缘山区的刚竹属等。

这些竹类植物生长环境多在海拔1500～3000米的高山地带。其中，分布地海拔最低的是镰序竹属，主产于喜马拉雅南麓及中国西南部（如岷山等山系），生长海拔450～1200米；分布海拔最高的是箭竹属，其垂直分布在海拔1500～3800米的地带。

受自然环境影响，同一竹类在不同山系，生长海拔也存在差异。由于大熊猫对各类竹类的喜好不尽相同，在不同季节，大熊猫对喜食竹的选择也有所不同，其活动范围呈现出周期性垂直变化。大熊猫最佳的食物基地是竹类繁多、出笋丰富的区域。

主食竹的分布区域又与山地型森林地带相互重叠。不同植被和竹类互利共生，它们与其他植物一起形成了稳定的森林群落。其中最主要的竹类群落分为箭竹群落、玉山竹群落、方竹群落和巴山木竹群落四大类别。这些群落不仅为大熊猫提供了隐蔽条件，还为它们提供了丰富的食物来源。

版画《进膳》，17.5cm×23cm（2018年）
敖正燊 作品

Chapter 3 Bamboos

Printmaking: *Enjoying Bamboos*, 17.5cm×23cm (2018). By Ao Zhengshen

In the giant panda's habitat, there are over 40 varieties of edible bamboos, among which 30 or more serve as the staple food. They are *Fargesia franchet emend.Yi* (arrow bamboo), the most widely distributed one; *Yushania Keng f*, mainly distributed in eastern Asia and Africa; *Indocalamus nakai*, along the Yangtze Valley and in mountainous and hilly regions of southern China; *Drepanostachyum*, in the southern slope of the Himalayas and southwestern China; *Bashania Keng f. et Yi*, in Sichuan, western Hubei, southern Shaanxi, and southern Gansu; *Qiongzhuea hsueh et Yi*, in southwestern China; *Chimonobambusa makino*, in China, Japan, Myanmar and India; and *Phyllostachys sieb.et Zucc*, in the mountainous region at the western edge of the Sichuan Basin.

Most of them grow in the highlands 1,500-3,000m above sea level, of which *Drepanostachyum* mainly grows at the southern foot of the Himalayas and southwestern China (Minshan) in areas of 450-1,200m in altitude while arrow bamboo grows in areas of 1,500-3,800m.

Bamboo of the same species may grow in areas of different altitudes in different mountains. As the giant panda has varied preferences to bamboos in different seasons, its scope of activity presents a cyclical and vertical change. Areas with various bamboo species and abundant bamboo shoots are the most ideal bases for the giant panda.

The distribution area of staple edible bamboo overlaps with the montane forest zone. Varieties of vegetation and bamboo exist symbiotically, thus forming a stable forest community. The major bamboo community can be categorized as the arrow bamboo community, *Yushania Keng f.* community, *Chimonobambusa quadrangularis* (Fenzi) Makino community and *Bashania fargesii* community, all of which make good shelters for the giant panda, thanks to their rich food resources.

青川箭竹、糙花箭竹、缺苞箭竹、华西箭竹（2012 年） 李仁贵 摄

Fargesia rufa, Fargesia scabrida, Fargesia denudata, Fargesia nitida (2012). Photo by Li Rengui

箭竹

大熊猫食用竹类中，箭竹属种类最多、分布最广，是竹亚科的一个大属，主要分布在中国西南高山地带，尤以横断山脉的种类最为丰富。

Fargesia franchet emend. Yi

Among the edible bamboos, *Fargesia franchet emend. Yi* has the largest kind species and widest distribution. It is a large genus in the *Bambusoideae*, mainly growing in the mountainous areas of southwest China, especially Hengduan Mountains.

绝技（2007 年） 周孟祺 摄

Stunt (2007). Photo by Zhou Mengqi

方竹

常集中成片地出现在山地沟谷常绿阔叶林或杉木林下，也有小片纯林。除中国西南地区外，日本也有分布。

Chimonobambusa quadrangularis (Fenzi) Makino

Patches of *Chimonobambusa quadrangularis (Fenzi) Makino* forest grow in valleys near the evergreen broad-leaved forests or Chinese fir forests. It is native to southwestern China and Japan.

巴山木竹

分布于北亚热带山地常绿阔叶林与落叶阔叶林中，如中国四川及湖北西部、陕西南部、甘肃南部等地。

Bashania fargesii

Bashania fargesii is distributed in the broad-leaved forests and evergreen forests in northern subtropical mountains, including Sichuan, western Hubei, southern Shaanxi and southern Gansu.

玉山竹

分布于亚洲东部及非洲，其中，绝大多数竹种产于中国南方高山地区。

Yushania Keng f.

Yushania Keng f. is distributed in eastern Asia and Africa, most of its varieties found in the high mountains in southern China.

大熊猫与小熊猫 引自魏辅文著《野生大熊猫科学探秘》，科学出版社，2018 年

The Giant Panda and The Red Panda. Cited from Wei Fuwen, *Scientific Exploration of Wild Giant Pandas*, Science Press, 2018

第四章 伙伴

大熊猫是动物界的“独行侠”，但在它的栖息地从来不缺少“朋友”。大熊猫分布的六大山系，为许多珍稀动物提供了绝佳的生存环境。

戴维在中国四川省穆坪（今雅安市宝兴县）科学发现大熊猫的同时，还发现了包括川金丝猴在内的数十种珍稀动物。随后一百余年时间里，通过对六大山系的调查和研究，人类揭开了这片密林中许多动物的神秘面纱。

其中较为著名的包括川金丝猴、扭角羚、朱鹮、小熊猫、毛冠鹿、金猫、大灵猫、小灵猫、红腹角雉等国家重点保护动物。这些跳动的精灵，打破了森林的寂静，把勃勃生机注入高山峡谷。

除了动物之外，六大山系也生长着不少被国家重点保护的珍稀植物，如珙桐、光叶珙桐、水青树、红豆杉、连香树、香果树、金钱槭、四川红杉、油麦吊云杉、大叶柳等树种。这些珍贵的动植物和大熊猫相伴，共同构成了丰富多彩的自然王国。

Chapter 4 Neighbors

The giant panda, a "loner" in the animal world, is never in short of playmates in its habitat. The six major mountain ranges provide a perfect living environment for various rare animals.

In 1869, Armand David discovered this species in Sichuan, China in addition to 10 other species of rare animals including the golden snub-nosed monkey. Over 100 years of investigation afterwards has made known many animals inhabiting the lush forests of the six major mountains.

Many species, such as golden snub-nosed monkey, takin, crested ibis, red panda, tufted deer, snow leopard, golden cat, zibet, rasse, tragopan temminckii and the like, are under special protection of the the State. These adorable creatures break the silence of the forests and bring vitality into the peaceful mountains and valleys.

In addition, there are also many species of rare plants, including dove tree, tetracentron, yew tree, *emmenoptery, (emmenoptery) henryi, Dipteronia sinensis, Larix mastersiana, Picea brachytyla,* and *Salix magnifica*. All these state-level rare flora and fauna, together with the giant panda, make a colorful and splendid natural paradise.

国画《锦绣天府 国宝家园》
360cm×145cm（2019 年） 吴长江 作品
Chinese painting: *Home to National Treasure*, 360cm×145cm (2019). By Wu Changjiang

油画《上树》，136cm×68cm（2019 年） 吴长江 作品

Oil Painting: *Climbing a Tree*,136cm×68cm (2019). By Wu Changjiang

川金丝猴（2016 年） 薛康 摄

Golden snub-nosed monkey (2016). Photo by Xue Kang

国画《松林啸月图》《中华精灵》《灵猴报瑞》，尺寸均为 140cm×65cm（2019 年） 胡真来 作品

Chinese paintings: *Pine Forest in the Moon*; *Chinese Elf*; *The Monkey: A Mascot of Good Fortune*, 140cm×65cm (2019). By Hu Zhenlai

川金丝猴

属猴科，仰鼻猴属，国家一级保护动物。体型中等，鼻孔上仰，蓝面黄毛，颊部及颈侧棕红。成年雄体川金丝猴身长 70 厘米左右，尾巴和身体几乎等长，所以当地人称它为“长尾巴猴”。

川金丝猴常年栖息于海拔 1500 ～ 3300 米的森林中。它的食性很杂，但以植物性食物为主，经常在林间采摘冬青、野樱桃、山楂、猕猴桃、拐枣、松子、板栗、高山栎、杜鹃花等作为食物。

Golden snub-nosed monkey

The golden snub-nosed monkey (*Rhinopithecus roxellana*), under first-class state protection, is of medium build with upturned nostrils, blue face, yellow hair, red-brown cheeks and brown neck sides. An adult male is about 70cm long, with a tail of almost the same length, hence the local people call it “long-tailed monkey.”

This kind of monkey lives in the forests of 1,500-3,300m above sea level as an omnivore, mainly feeding on plants such as holly, wild cherry, hawthorn, kiwi fruit, hovenia dulcis, pine nuts, chestnut, *quercus semecarpifolia smith, azaleas* and others.

扭角羚（1996 年） 周孟棋 摄
Takin (1996). Photo by Zhou Mengqi

扭角羚

属牛科，羚牛属，国家一级保护动物。体型粗大，体长为 1.7 ～ 2.2 米，体重可达 400 千克。扭角羚均具较短的角，角呈扭曲状，一般长约 20 厘米，又称羚牛。扭角羚通常栖息在海拔 1600 ～ 4000 米的山林，随着季节和食物基地的变化，每年定期上山下山做有规律活动。

Takin

The Takin (*Budorcas taxicolor*), under first-class state protection, is of bulky build, around 1.7-2.2m in length, and 400kg in weight, with short and twisted horns of about 20cm in length. It usually inhabits the forest at an altitude of 1,600-4,000m above sea level, migrating with the seasonal change up and down the mountains searching for edible plants.

小熊猫（2012 年） 黄耀华 摄

Red Panda (2012). Photo by Huang Yaohua

小熊猫

属小熊猫科，小熊猫属，国家二级保护动物。它的身体长度在 60 厘米左右，躯体肥壮，尾巴又长又粗，身上的长毛呈红褐色，体重约 5 千克。小熊猫主要在海拔 3000 米以下的针阔混交林或常绿阔叶林中有竹丛的地方生活，平时在树洞或石洞、岩石缝中栖居。小熊猫在春天喜欢吃竹笋，其余时间以竹叶为食，在秋天也采食少量浆果类食物。

Red Panda

The Red panda (*Ailurus fulgens*), under second-class state protection, has a reddish-brown fur and long, shaggy tail with its body around 60cm in length, and an average weight of 5kg, mainly inhabiting the mixed conifer-broadleaved and evergreen broadleaved forests below 3,000m in altitude. It dwells in a tree hollow, stone cave or rock crack, and is fond of bamboo shoots in spring, bamboo leaves in other seasons, in addition to a small amounts of berry in autumn.

珙桐

落叶乔木，中国特有的单属植物，属孑遗植物，是国家一级保护野生植物，也是全世界著名的观赏植物，被誉为“中国的鸽子树”。珙桐一般能长到 15 ～ 20 米，少数能达到 25 米，叶子呈广卵形，边缘有锯齿；树皮呈深灰色或深褐色，常裂成不规则的薄片而脱落。珙桐喜欢生长在海拔 1500 ～ 2200 米的常绿阔叶、落叶阔叶混交林中，野生种只生长在中国西南部的四川省、中部的湖北省和周边地区。

珙桐花（2006 年） 王达军 摄

Dove tree in bloom (2006). Photo by Wang Dajun

Dove Tree

Deciduous dove trees (*Davidia involucrate*), a monotypic genera plant peculiar in China (hence nicknamed "Chinese dove trees"), are relic plants under the First-Class State Protection of Plants. As its flower looks like a flying dove, it is favored as an ornamental plant worldwide; it grows to be 15-20m tall, with a few up to 25m, in evergreen broadleaved and deciduous broadleaved forests of 1,500-2,200m in altitude in Sichuan, Hubei, and surrounding provinces. Its leaf is oval-shaped and serrate-edged, and its bark is likely to flake off.

第四篇
研究

自贡扎染《竹与大熊猫》，110cm×90cm（20 世纪 90 年代） 自贡市地方志办公室 供图

Zi Gong Tie-dye: *Bamboo and the Giant Panda*, 110cm×90cm (1990s). Photo courtesy of Zigong Local Records Office

Part 4
Research

好奇之光闪烁，照亮神秘之地

生命的幽微细节，

在两个物种的相遇中显现。

大熊猫递给人类一把打开生物演化奥秘的钥匙，

人类为大熊猫推开了种族延续之门。

The light of curiosity illuminates this mysterious land

The subtleties of life are revealed

in the encounter between the giant panda and man.

The giant panda offers man an insight into the mystery of biological evolution,

while man creates new chances for the giant panda to prolong its existence on the planet.

双桥沟风光（2016 年） 周孟棋 摄

Shuangqiaogou landscape (2016). Photo by Zhou Mengqi

第一章 生态

自1869年以来，世界生物学界对大熊猫这一迷人物种的研究热情与日俱增。

20世纪80年代初，在中国政府与世界野生动植物基金会（WWF）的共同推动下，中外50多名专家在卧龙五一棚、唐家河白熊坪对野生大熊猫进行了全方位的观测和研究。经过2年多的调研，研究界逐步揭开了野生大熊猫生活习性的面纱。

与此同时，岷山、邛崃山等地竹子开花枯死的灾害爆发，让大熊猫研究受到了更广泛关注，研究力量随之增强，研究成果也越来越丰硕。众多研究成果当中，营养生态学研究、种群生态学研究、分子生态学研究、种群遗传学研究等成果既有自主创新，又有国际合作，让大熊猫研究步入全新阶段。

近年来，科学家们开始通过更前沿的科学手段对大熊猫进行研究。比如，通过无线电追踪，了解其活动规律和迁移扩散；通过粪便DNA分析法，进行野外大熊猫样本的分子遗传学分析，评估野生大熊猫遗传多样性水平等。

科学家们的积极探索为大熊猫研究提供了更多思路。

国画《九寨清韵》，240cm×143cm（2018年） 任伟 作品
Chinese painting: *Vivacious Jiuzhaigou*, 240cm×143cm (2018). By Ren Wei

Chapter 1 Ecology

国画《熊猫格格》，138cm×68cm（2018 年） 沙马拉毅 作品

Chinese painting: *Giant Panda Ge Ge*, 138cm×68cm (2018). By Shama Layi

Ever since 1869, biologists around the world have been increasingly interested in the study of this fascinating species—the giant panda.

Early in the 1980s, co-sponsored by the Chinese Government and World Wide Fund for Nature (WWF), more than 50 experts from China and other countries made an all-round observation of and research on the wild giant panda at Wuyipeng Observation Station of Wolong National Nature Reserve and Baixiongping of Tangjiahe Nature Reserve. Over two years of research gradually revealed its living habits.

Meanwhile, the disastrous bamboo flowering in such areas as Minshan Mountains and Qionglai Mountains attracted attention to the research of the giant panda, resulting in better research teams and more fruitful research findings. Some findings were made by the Chinese scientists while others were the results of international cooperation; the findings touched upon nutrition ecology, status quo of existence, molecular biology and genetic diversity. From then on, research on the giant panda has entered a new stage.

In recent years, cutting-edge scientific technologies have been applied by scientists to the study of the giant panda. For example, radio transmitters have been employed to track its activity and mobility; faecal DNA has provided for the analysis of the wild giant panda samples concerning molecular genetics and genetic diversity.

The scientists' persistent efforts have provided them with new approaches toward conservation.

成都大熊猫繁育研究基地的大熊猫“奥莉奥”在吃竹笋（2018 年）
黄馨 摄

Giant panda Oreo is crunching bamboo shoots at Chengdu Research Base of Giant Panda Breeding (2018). Photo by Huang Xin

营养学研究

大熊猫营养学研究以大熊猫主要食物——竹类食物及少量其他食物与机体的互相作用，以及食物营养成分在机体里分布、运输、消化和代谢等为主要研究内容。也可理解为食物中的营养素和其他物质间的互相作用与平衡，对大熊猫健康和疾病的关系，以及大熊猫摄食、消化、吸收、转运、利用、排泄物质的过程。

Research on Nutriology

Nutriology focuses on the interaction between the giant panda's staple food (bamboos and small quantities of other foods) and its organs, as well as distribution, transportation, digestion and metabolism of nutrients within its body. In other words, nutriology deals with how the panda's health is influenced by the interaction and balance between nutrients and other substances in food, and how it takes in, digests, absorbs, transports nutrients and excretes waste.

右 涉水（2008 年） 周孟棋 摄

(Right) Swimming across the river (2008). Photo by Zhou Mengqi

尝试爬树（2008 年） 周孟棋 摄

Trying to climb the tree (2008). Photo by Zhou Mengqi

生存态势研究

生存态势研究注重研究大熊猫的种群结构。20 世纪 90 年代，科学家通过调查发现，中国四川省境内的野生大熊猫年龄结构为金字塔形，具有缓慢增长的潜力。只要大熊猫的雌雄比例合理，它们的种群就能保持总体上的稳定。避免野生大熊猫近亲繁殖，加强灾害预防，可以促进种群的长期繁衍。

Research on Survival Situation

The structure of the giant panda population is the focus of the research. In the 1990s, scientists found that the wild giant panda population in Sichuan has the potential for a gradual increase, because its age structure is pyramidal. So as long as the male-female ratio is well-balanced, its population will remain stable on the whole. Inbreeding prevention among the wild giant panda, and circumvention of likely disasters can promote its long-term reproduction.

我也来看看监控 引自魏辅文著《野生大熊猫科学探秘》，科学出版社，2018 年
Let me check on the camera. Cited from Wei Fuwen, *Scientific Exploration of Wild Giant Pandas*, Science Press, 2018

遗传多样性研究

遗传多样性，指种内个体之间或一个群体内不同个体的遗传变异总和。中国的大熊猫遗传保护技术已处于世界领先地位。近年来，中国科学家通过分子粪便学和大规模野生大熊猫样本的遗传变异分析，重新评估了野生大熊猫的遗传多样性水平，无论是野生大熊猫或是圈养大熊猫，无论是线粒体分子标记还是微卫星及基因组单核苷酸分子标记，均表现出较高的遗传多样性。

Research on Genetic Diversity

Genetic diversity refers to the total number of genetic variations among individuals within a species or a population. In recent years, some Chinese scientists have been applying molecular scatology to the study of large quantities of faecal samples from the wild with the intent of reevaluating the wild giant panda's genetic diversity. A high genetic diversity has been found in both the wild and captive giant panda, as has been indicated by mitochondrial DNA molecular markers, microsatellite markers and single nucleotide polymorphism markers.

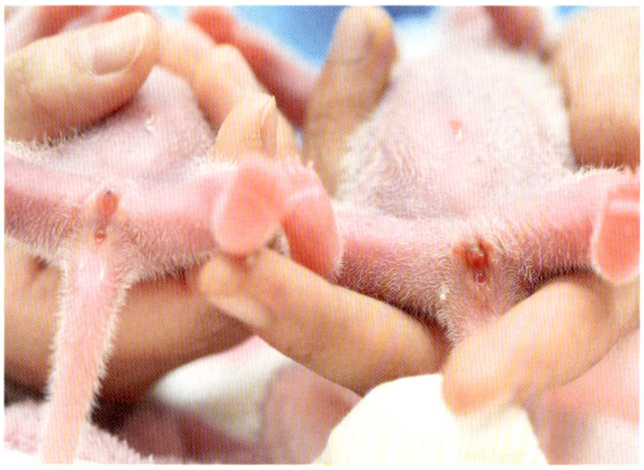

雌雄差异主要体现在会阴部隆起部至肛门的间距，左为雌体，间距短；右为雄体，间距长（2011 年） 佘轶 摄

The distance between perineal eminence and anus is short for female (left), and long for male (right) (2011). Photo by She Yi

性别鉴定研究

过去，鉴定大熊猫性别较为困难。随着大熊猫科学研究水平的提高，对大熊猫的性别鉴定现已成为一种成熟的技术。富有经验的专家能通过幼崽生殖器至肛门间的距离来判断它的性别。此外，随着分子粪便学的发展，科学家们已经能够准确鉴定野生大熊猫的性别、调查野生大熊猫的性别结构。

Research on Sex Identification

With the development of scientific research, sex identification has become a mature technology for the giant panda though difficult in the past. Experienced experts can tell it from the distance of the genital to anus. And, with the development of molecular scatology, they can investigate its sex structure.

右 拍好了吗？（2012 年） 杨老丹 摄

(Right) Posing for pictures taken (2012). Photo by Yang Laodan

春笋破土 引自四川省林业厅编《图说大熊猫》，天地出版社，2015 年
Spring bamboo shoots push up through the soil. Cited from *The Giant Panda Pictures* compiled by Forestry Department of Sichuan Province and published by Tiandi Press, 2015

剪纸《熊宝趣味图》，27cm×21cm（2017 年）
杜华江 作品

Paper-cutting: *Happy Cub*,27cm×21cm (2017).
By Du Huajiang

剪纸《熊猫》，四开（20 世纪 90 年代） 叶牧天 作品

Paper-cutting: *The Giant Panda*,Quarto (1990s). By Ye Mutian

第二章 食物

大熊猫的主食是竹子，当竹子大面积开花枯死后，大熊猫容易因缺少食物而死亡。

为解决竹子开花危机，科研工作者们通过 10 多年的研究，掌握了主食竹更新复壮、延期开花、人工栽植、引种栽培等技术，力图让大熊猫时刻有鲜嫩的竹子可食用，从根本上解决大熊猫缺食危机。也正是这 10 多年，专家们填补了国内外对高山竹类研究的多项空白。

除此之外，科研工作者们还对竹子开花原因、主食竹开花预测与观察、主食竹开花与大熊猫繁衍、主食竹保护与恢复等方面进行了具体研究。有专家认为，大熊猫从古至今都面临着竹子开花后的缺食危机，却没有灭绝，所以竹子开花并不是阻碍大熊猫繁衍的最终原因。

通过大量的观察与试验发现，保护森林资源能提高竹子质量和分布密度，让主食竹得到更好生长。所以在保护大熊猫的过程中，人们试图采取人工手段来保护主食竹，恢复森林生态结构，为大熊猫创造舒适的生存环境。

左 / 右 铅笔画《闹雪》，56cm×76cm（2019 年） 薛霞 作品

(Left/Right) Pencil sketch: *Play in the Snow*, 56cm×76cm (2019). By Xue Xia

Chapter 2 Staple Food

If bamboo, as the staple food, flowers and dies extensively, the giant panda will most probably be doomed from food shortage.

It has taken researchers more than ten years to solve this food crisis. The solutions lie in the renewal and rejuvenation of fodder bamboo, including delayed flowering, bamboo planting, and introduction of new varieties of bamboo. In the recent decade, the experts have filled in many gaps in the high-alpine bamboo research.

Also, researchers have pursued the causes for bamboo flowering, made predictions of the flowering season, studied the effect of bamboo flowering on the giant panda propagation, and provided some measures for the conservation of staple edible bamboo. Some experts believe that bamboo flowering is not the critical factor that has the hindered giant panda's reproduction, because it has survived bamboo flowering ever since ancient times.

Recurrent observations and experiments have found that the protection of forest resources can improve the quality and distribution density of fodder bamboo. Therefore, in the endeavors for giant panda conservation, human intervention can help create a livable environment.

竹子开花 引自四川省档案局、四川省档案学校编《档案见证·大熊猫》，四川人民出版社，2018 年

Bamboo flowering. Cited from *Archives Witness: The Giant Panda* compiled by Sichuan Province Archives Bureau and Sichuan Archives College, and published by Sichuan People's Publishing House, 2018

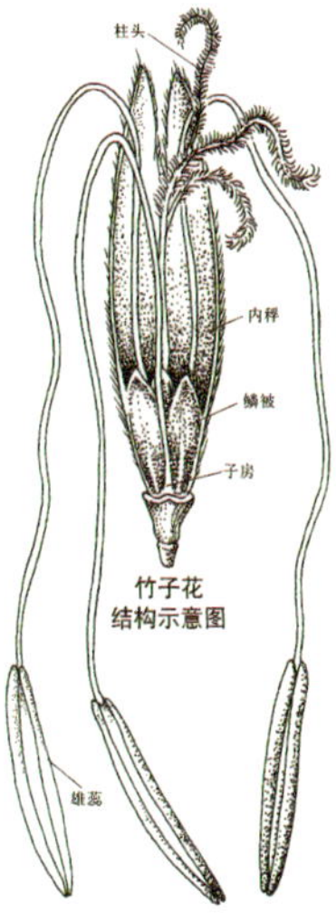

左 竹子花结构示意图 引自赵学敏编《大熊猫——人类共有的自然遗产》，中国林业出版社，2006 年

右 1983 年，卧龙国家级自然保护区竹子大面积开花，大熊猫面临着死亡威胁 新华社 供图

(Left) A diagram of bamboo flower structure. Cited from Zhao Xuemin, *The Giant Panda: Natural Heritage of Humanity*, China Forestry Publishing House, 2006

(Right) Giant pandas faced the threat of death as most bamboos in Wolong National Nature Reserve in Sichuan Province flowered in 1983. Photo courtesy of Xinhua News Agency

竹子开花

开花是植物生长过程中正常的生命现象。竹子的开花结实是利用有性生殖方式，完成竹类的更新与延续。其原因主要有两种说法：其一认为竹子开花是一种正常的生理现象，标志着竹株的成熟和衰老；其二认为是由于外部环境条件变化导致竹子开花。

部分竹子只开花一次，而且有一定周期；少数竹子年年开花，花后竹竿不会死亡，仍可抽鞭发笋；还有一类竹子不定期零星开花，花后竹竿也不会死去。但大熊猫喜好采食的箭竹生长成熟后，只要开花结实，地上地下整个植株都会死亡。

Bamboo Flowering

Flowering is a normal phenomenon for the growth of plants. Bamboo flowering and fruiting is programmed to complete the renewal and continuation of this plant through sexual reproduction. Two explanations are proposed to account for this phenomenon. One goes that bamboo flowering is a normal physiological phenomenon, marking maturity and ageing of bamboo, and the other says that it results from changes in external environments.

Some species of bamboos flower only once in their lifetime and die after that. A small number of them flower every year, and they shoot up rather than die soon. There is still another variety of bamboo that flowers aperiodically and sporadically, but it remains alive after blossoming. *Fargesia apathacea* (arrow bamboos)—the giant panda's favorite food, will die once flowering.

五一棚研究人员踏勘大熊猫主食竹 引自四川省地方志编纂委员会编《四川省志·大熊猫志》，方志出版社，2018 年

The researchers of Wuyipeng Observation Station are making an on-the-spot survey of the fodder bamboos of giant pandas.Cited from *Sichuan Province Annals: Giant Panda* compiled by Sichuan Local Records Editorial Committee, and published by Publishing House of Local Records, 2018

光谱观测

不同种的主食竹以及同一种主食竹在不同季节、不同生长周期，其光谱反射率均不同。通过光谱变化可观察竹子的生长状况，比如愈加临近主食竹的开花时节，其近红外波段光谱反射率愈加降低。

Spectral Observation

Varieties of staple bamboo differ in spectral reflectance rates, as does the same variety of staple edible bamboo in different seasons and periods of growth. The growth performance of bamboo can be observed through spectral changes. For example, as bamboo approaches its flowering season, its near-infrared wave spectral reflectance rate decreases.

在四川省崇州市鞍子河镇境内的大熊猫（2018 年） 鞍子河国家级自然保护区 供图

A giant panda in Anzihe Township, Chongzhou City, Sichuan Province (2018). Photo courtesy of Anzihe National Nature Reserve

竹子开花与大熊猫进化

有学者认为，箭竹开花枯死，促进了大熊猫的优胜劣汰。优良的个体幸存下来，病弱的则被淘汰。同时，为了寻找新的食源，大熊猫不得不迁移异地，这有利于大熊猫远缘杂交，避免近亲结合。枯死的老竹化为营养成分滋养新生竹林，为大熊猫提供了新的食物基础。

Bamboo Flowering and Giant Panda's Evolution

Some scholars believe that the flowering and death of arrow bamboo promotes the selective evolution of the giant panda, by which the strong flourish while the weak perish. At the same time, to find new sources of food, some of the giant panda have to move on to other areas, which can promote different lineage mating and avoid inbreeding. The dead bamboo turns into nutrients, nourishing the newborn and providing a new food base for the giant panda.

照镜子（2015 年） 周孟棋 摄

Looking in the mirror (2015). Photo by Zhou Mengqi

漫步（2008 年） 周孟祺 摄

Strolling by the river (2008). Photo by Zhou Mengqi

大熊猫嗜水

中国科学院成都地理研究所（今中国科学院水利部成都山地灾害与环境研究所）研究员郑远昌研究发现，卧龙国家级自然保护区内的大熊猫愿意喝离子含量每升在 80 ～ 265 毫克之间的中等硬度溪水，不喝离子含量低的极软水，也不喝离子含量高的硬水。针对大熊猫的爱好，郑远昌发现该保护区的水源十分得大熊猫的“芳心”。

Giant Panda’s Favorite Type of Drinking Water

Zheng Yuanchang, a researcher at Chengdu Institute of Geography, Chinese Academy of Science [now known as Institute of Mountain Hazards and Environment (IMHE), Chinese Academy of Science, Chengdu], found that the giant panda in Wolong National Nature Reserve prefer medium-hard stream water with an ion content of 80-265mg per liter, as it drinks neither extremely soft water with a low ion content nor hard water with a high ion content. Based on this, Zheng found that the stream water in Wolong Nature Reserve is a favorite with it.

2004 年，科研工作者为大熊猫做电刺激采精 成都大熊猫繁育研究基地 供图
Collecting sperms by electrical stimulation in 2004. Photo courtesy of Chengdu Research Base of Giant Panda Breeding

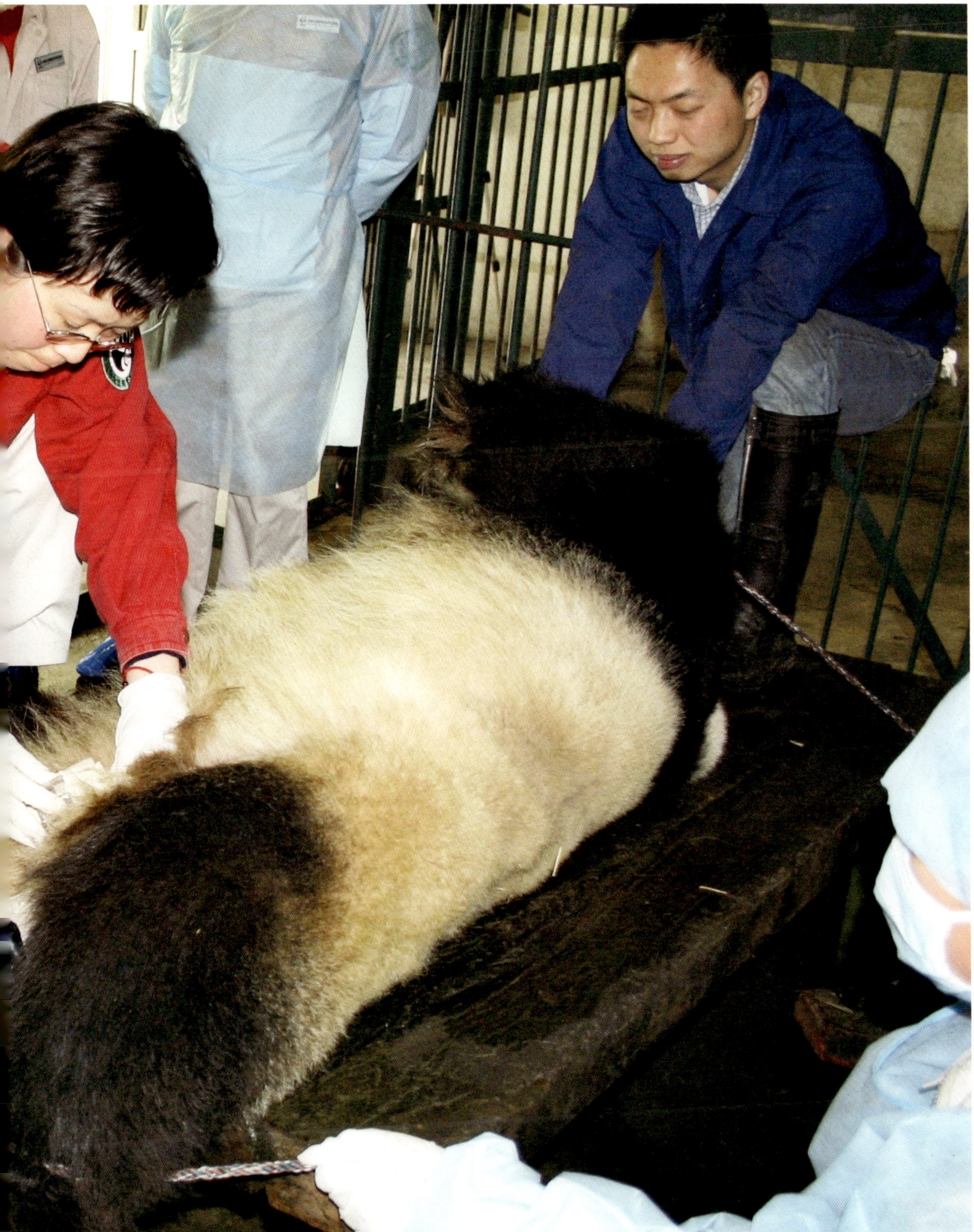

第三章 圈养繁殖

1963 年，北京动物园首次利用具有自然交配能力的雄性大熊猫“森森”和雌性大熊猫“莉莉”进行自然交配，成功产下一崽。这是世界上第一例人工繁殖的大熊猫，从此拉开了人工繁殖大熊猫的序幕。但是 1963—1977 年，人工繁育大熊猫的科学技术尚不成熟，圈养种群主要是野外捕捉和抢救的大熊猫进行自然交配而产下的后代。1978 年后，研究者们加大了对圈养大熊猫人工繁殖技术的攻坚力度。

北京动物园在 1978 年采用电激法对 4 只大熊猫进行了人工授精，其中有 1 只大熊猫产下双胞胎。这是世界上第一例人工授精繁殖大熊猫的成功案例。

1980 年，成都动物园利用冷冻精液人工授精繁殖大熊猫获得成功，为解决圈养大熊猫发情难、配种受孕难和育幼存活难问题（以下简称“三难”问题）带来了希望。

1997 年，全国范围内圈养出生的大熊猫个体数量首次超过野外批准获取的大熊猫个体数量，并且每年以 12% 的速度增长。1998 年后，中国保护大熊猫研究中心的大熊猫人工繁育成活率超过 80%，标志着 20 世纪 90 年代末，中国成功攻克圈养大熊猫“三难”问题。成都大熊猫繁育研究基地也为大熊猫人工繁育研究做出了突出贡献。该机构以 20 世纪 80 年代抢救留下的 6 只病饿大熊猫为基础，在未从野外捕获大熊猫的情况下，截至 2018 年，共繁育大熊猫 188 胎、284 只，现存 195 只，是全球最大的圈养大熊猫人工繁殖种群之一，且种群遗传质量、种群内个体健康状况和行为健康状况均良好。

2018 年，共繁殖大熊猫 36 胎、48 只，存活 45 只。截至 2018 年底，中国圈养大熊猫种群数量再创新高，全球圈养数量达到 548 只。大熊猫已基本达到种群自我维持的状态，为复壮野生种群奠定了坚实基础。

油画《青花瓷·伊甸园·立秋》
60cm×80cm（2017 年） 谢纲骅 作品
Oil painting: *Start of Autumn in the Garden of Eden on Chinese Porcelain*,60cm×80cm (2017). By Xie Ganghua

成都动物园的大熊猫“庆庆”，1989 年 5 月通过人工授精后，于 10 月 4 日喜产贵崽“星星”。图为“庆庆”和“星星” 新华社 供图

Qing Qing, which received artificial insemination in Chengdu Zoo in May 1989, birthed a cub on October 4, 1989. The cub named Xing Xing is now in its mother’s arms. Photo courtesy of Xinhua News Agency

1

2

1. 插画《中药制作：炒》，38cm×26cm（2018 年） 成都市实验小学 郭子瑜、马羽杉 作品 指导老师 王梅
2. 插画《中药制作：捣》，38cm×26cm（2018 年） 成都市实验小学 杨若熙、钟瑗如 谢紫轩 作品 指导老师 王梅
3. 插画《中药制作：碾》，38cm×26cm（2018 年） 成都市实验小学 周千千、马羽杉 作品 指导老师 王梅
4. 插画《果宝教你文明骑行》，38cm×26cm（2017 年） 成都市实验小学 姚果果 作品 指导老师 罗东来

Chapter 3 Captive Breeding

In 1963, a giant panda cub was born in Beijing Zoo as the result of the natural mating of Sen Sen (male) and Li Li (female). This was the world's first born in captivity, which kicked off the giant panda's artificial breeding. From 1963 to 1977, the technology of giant panda's artificial breeding was not effective enough. The main captive population was made up of the cubs birthed through natural mating by the giant panda captured or rescued from the wild. Since 1978, researchers have put in greater efforts on artificial reproduction of the captive giant panda.

In 1978, Beijing Zoo conducted an artificial insemination through electro-stimulating four giant pandas and one of them gave birth to a pair of cubs. This was the first success of its kind in the world.

In 1980, Chengdu Zoo succeeded in applying frozen semen to inseminate estrous female pandas, heralding the solutions to the "three difficulties" in breeding: getting the animals estrous, pregnant, and caring for the cub.

In 1997, for the first time in history, baby pandas born in captivity outnumbered those captured from

3

4

1.Illustration: *Traditional Chinese Medicine: frying*, 38cm×26cm (2018). By Guo Ziyu and Ma Yushan from Chengdu Experimental Primary School, under the instruction of Wang Mei

2.Illustration: *Traditional Chinese Medicine: Pounding*, 38cm×26cm (2018). By Yang Ruoxi, Zhong Airu and Xie Zixuan from Chengdu Experimental Primary School, under the instruction of Wang Mei

3.Illustration: *Traditional Chinese Medicine: Grinding*, 38cm×26cm (2018). By Zhou Qianqian and Ma Yushan from Chengdu Experimental Primary School, under the instruction of Wang Mei

4.Illustration: *Civilized Parking*, 38cm×26cm (2017). By Yao Guoguo from Chengdu Experimental Primary School, under the instruction of Luo Donglai

the wild, with a 12% annual increase from then on. Since 1998, the survival rate of cubs born at China Panda Center has exceeded 80%, indicating that China circumvented the three difficulties in the 1990s. Chengdu Panda Base has made significant contributions to panda artificial breeding. Starting from 6 sick and starving individuals rescued from the wild in the 1980s, the Base has bred 284 cubs in 188 litters, 195 of which survived (as of 2018). These captive giant pandas are one of the largest artificial breeding populations in the world, possessing an exceptional genetic quality and good health physically and behaviorally.

In 2018, 45 cubs survived out of 48 in 36 litters. By the end of 2018, the population of captive giant pandas in China had hit a new high, while the total number of captive giant pandas reached 548 all over the world. Now, its population can be self-sustainable, which is significant for the rejuvenation of the wild giant panda population.

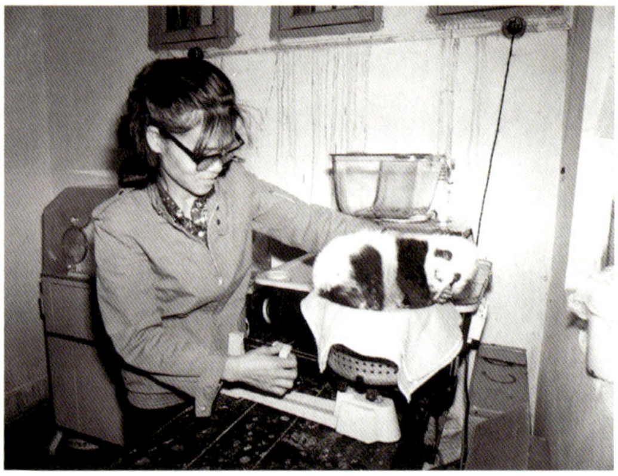

1986 年 8 月 6 日，成都动物园大熊猫“果果”，经人工授精产下一崽，取名“冰冰” 新华社 供图

On August 6, 1986, giant panda Guo Guo, which received artificial insemination, birthed a baby in Chengdu Zoo. The baby was named Bing Bing. Photo courtesy of Xinhua News Agency

科学喂养法

科学喂养可提高大熊猫的繁殖率。比如利用大熊猫吃剩下的竹子，将之粉碎后做成高纤维窝头和饼干，使大熊猫的消化系统能更好更快地消化食物。同时，增加大熊猫饲料的多样性和丰富性，可保证大熊猫的身体健康。

Scientific Feeding

Scientific feeding can improve the reproduction rate of the giant panda. As we know, when enjoying its delicious food, it leaves a lot (of leaves, stems and shoots) unconsumed. In order to make the best of the leftovers, keepers crush and make them into bread or biscuits, resulting in the fodder being high in fiber, and easy to digest.

右 护养（2016 年） 周孟棋 摄

(Right) Nursing (2016). Photo by Zhou Mengqi

2004 年，成都大熊猫繁育研究基地建立全球首个大熊猫基因资源库 成都大熊猫繁育研究基地 供图

Chengdu Research Base of Giant Panda Breeding established the first giant panda gene bank in the world in 2004. Photo courtesy of Chengdu Research Base of Giant Panda Breeding

大熊猫精子库

截至 2017 年底，成都大熊猫繁育研究基地大熊猫精子库共储存 42 只大熊猫个体（其中 16 只个体已死亡）的细管冻精 6391 支，颗粒冻精 1484 颗，是全球最大的大熊猫精子库。

通过建立精子库、细胞库等类型的基因资源库，可将大熊猫的遗传与生殖资源长久保存下来。同时，这些技术可为大熊猫就地和迁地保护做资源储备，进一步推动大熊猫的保护研究发展。

Giant Panda Sperm Bank

By the end of 2017, the sperm bank of Chengdu Research Base of Giant Panda Breeding had 6,391 frozen semen straws and 1,484 grains of frozen semen from 42 giant pandas (16 of which had died), becoming the largest panda sperm bank in the world.

The genome resource banks, including the sperm bank and cell bank, make it possible to preserve its genetic and reproductive resources in the long run while serving as a resource basis for its ex-situ and in-situ conservation, and further promoting the research and development of its conservation.

吃奶瓶的大熊猫（2007 年） 张志和 摄

Sucking from the bottle (2007). Photo by Zhang Zhihe

人工配方乳

幼崽人工配方乳的成功研制，使大熊猫的人工育幼成活率快速提高到 90% 以上。人工配方乳易于消化吸收，不仅可刺激幼崽免疫系统的发育，还同步攻破了亚成体大熊猫死亡率高、生长发育不良等难题。

Formula Milk

With the successful development of formula milk, the panda cub's survival rate in captivity has gone beyond 90%. Easy to digest and absorb, formula milk can stimulate its immune system as one of the solutions to such problems as high mortality rate and maldevelopment of sub-adults.

你在我眼中（2016 年） 周孟棋 摄
You're in my eyes (2016). Photo by Zhou Mengqi

第四章 疾病

在野外，一只健康的青壮年大熊猫从不畏惧豺狼虎豹，但它却容易受到寄生虫的伤害。

寄生虫病是大熊猫最常见的疾病，且体内、外都可能感染。圈养的大熊猫基本上不会因寄生虫病而死亡，但很多野生大熊猫都不同程度地被寄生虫感染。

除了寄生虫病，还有各种其他疾病困扰着大熊猫。如狂犬病、犬瘟热病、细小病毒、冠状病毒等传染性疾病。犬瘟热病毒在犬科动物群体中，可通过飞沫、病患的分泌物与排泄物等介质传播，这种疾病对大熊猫威胁极大。

尽管医道艰难，但为心爱的大熊猫与疾病斗争，是人类心之所向。在大熊猫疾病与病理研究的这条道路上，人类从未停止探索。

大熊猫疾病的研究最早可追溯到 1939 年，美国学者麦金托什发表了一篇名为《熊猫蛔虫新种》的论文。中国最早的大熊猫疾病研究是在民国 30 年（1941 年），“国立中央研究院动植物研究所”（今中国科学院水生生物研究所）的刘建康和伍献文首次在大熊猫体内发现了犬钩口线虫。

1978 年以来，大熊猫疾病和病理研究也取得了许多成果。曾参与过抢救野外受灾大熊猫医疗工作的叶志勇医师，对医疗抢救的 50 例野外大熊猫的疾病做了较为详细的分类统计，这份资料成了研究大熊猫疾病预防和救治的宝贵素材。

进入 21 世纪，科研人员对大熊猫疾病有了更深入的认识。2003 年 1 月，中国保护大熊猫研究中心编纂出版了《大熊猫繁殖研究》一书，基本形成了“大熊猫疾病谱”雏形，将大熊猫疾病防治提高到了新的水平。

高处不胜寒（2008 年） 谢浩 摄

Chapter 4 Diseases

gh up in the tree (2008). Photo by Xie Hao

Although a strong giant panda never fears any challenger in the wild, it is vulnerable to parasites.

The parasitic disease, as the most common ailment for the giant panda, can affect it both internally and externally. The captive giant panda rarely dies of it, but few wild giant pandas are free from such a disease.

Besides the parasitic disease, a variety of others afflict it as well, such as hydrophobia, canine distemper, parvovirus and coronavirus. Canine distemper virus in canidae can be spread through spittle, secretions, feces, etc, posing a great threat.

Thankfully, human beings have long been devoted to fighting diseases for their beloved giant panda and will never stop their pathological inquiries.

Research into panda diseases goes back to 1939, when Mcintosh, an American scientist, published a paper titled *A New Species of Ascaris Schroederi*. The first study of giant panda disease in China took place in 1941, when Liu Jiankang and Wu Xianwen from the Institute of Zoology and Botany of Academia Sinica (now known as the Institute of Hydrobiology of CAS), discovered ancylostoma caninum (commonly known as hookworm) in the giant panda for the first time.

Since 1978, many achievements have been made in the study of diseases and pathology afflicting the giant panda. Doctor Ye Zhiyong, who participated in the rescue of the wild giant panda that suffered starvation after the extensive bamboo flowering, made a detailed classification of the 50 diseases that affected the wild giant panda in 1987. This document has since been a vital reference for scientists dealing with such diseases.

In the 21st century, researchers have gained a still-deeper understanding of panda diseases. In January 2003, China Panda Center published the treatise of *A Reproductive Research of the Giant Panda*, which presents the basic "disease spectrum of the giant panda" and is helpful in the prevention and treatment of such diseases.

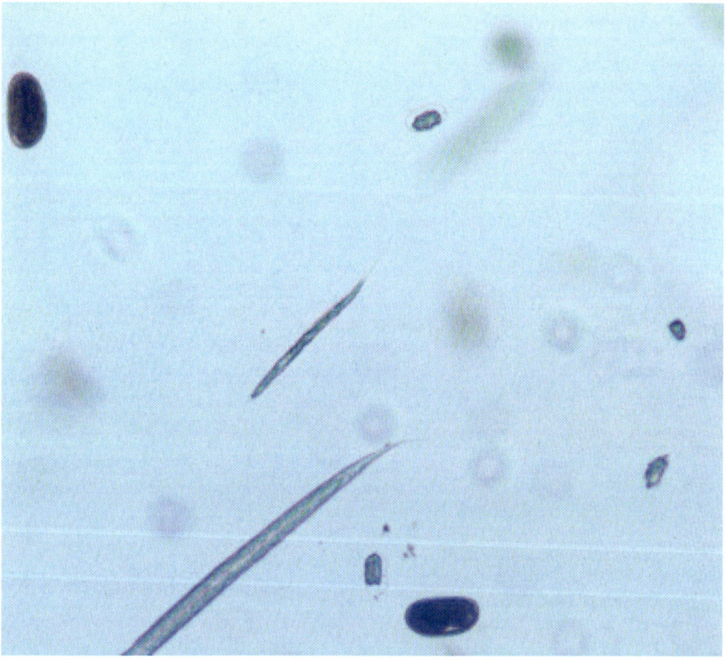
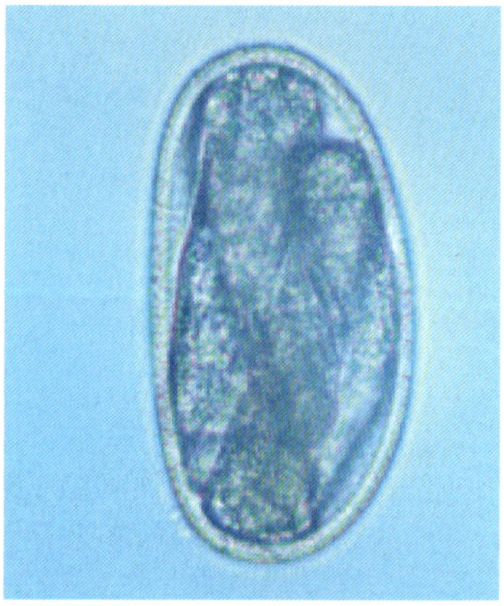

左／右上 大熊猫粪便内的钩虫卵 引自四川省档案局、四川省档案学校编《档案见证·大熊猫》，四川人民出版社，2018 年
右下 大熊猫的粪便 乐山市沙湾区地方志办公室 供图

(Left/Upper right) The hookworm eggs in giant panda's feces. Cited from *Archives Witness: The Giant Panda* compiled by Sichuan Province Archives Bureau and Sichuan Archives College, and published by Sichuan People's Publishing House in 2018

(Lower right) Giant panda's feces. Photo courtesy of Shawan District Local Records Office of Leshan City

蛔虫病

大熊猫感染蛔虫病与其生活习性有关。大熊猫食竹量大，所以粪便排量也大，而它们的活动范围又相对固定。蛔虫卵随粪便一起排在大熊猫的休息区周围，当大熊猫再次进食时很容易“病从口入”。

Ascariasis

The giant panda's infection of ascariasis is related to its living habits by eating a large amount of bamboo and defecating a great deal of feces within a close range of its activity. As a result, ascaris eggs are spread and contaminate the bamboo it eats.

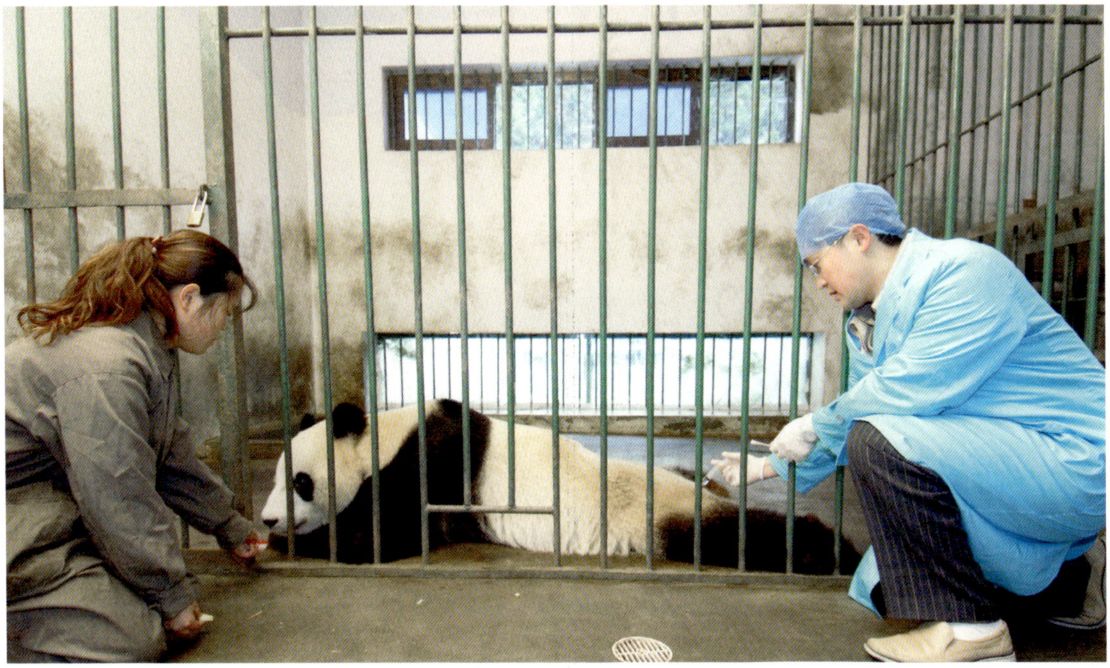

兽医正给生病的大熊猫打针 引自四川省档案局、四川省档案学校编《档案见证·大熊猫》，四川人民出版社，2018 年

Veterinarians are giving the sick giant panda an injection. Cited from *Archives Witness: The Giant Panda* compiled by Sichuan Province Archives Bureau and Sichuan Archives College, and published by Sichuan People's Publishing House, 2018

犬瘟热病

由犬瘟热病毒引起的一种以呼吸、消化和神经系统病变为主的急性、接触性、败血性传染病。大熊猫和小熊猫均高度易感，体弱或老年大熊猫对其敏感性更强，这种疾病致死率很高。

Canine Distemper

This viral infection is an acute, contagious and septic disease, fatally affecting the respiratory, digestive and nervous systems. Both the giant and red pandas are highly susceptible, especially the aged and weak ones.

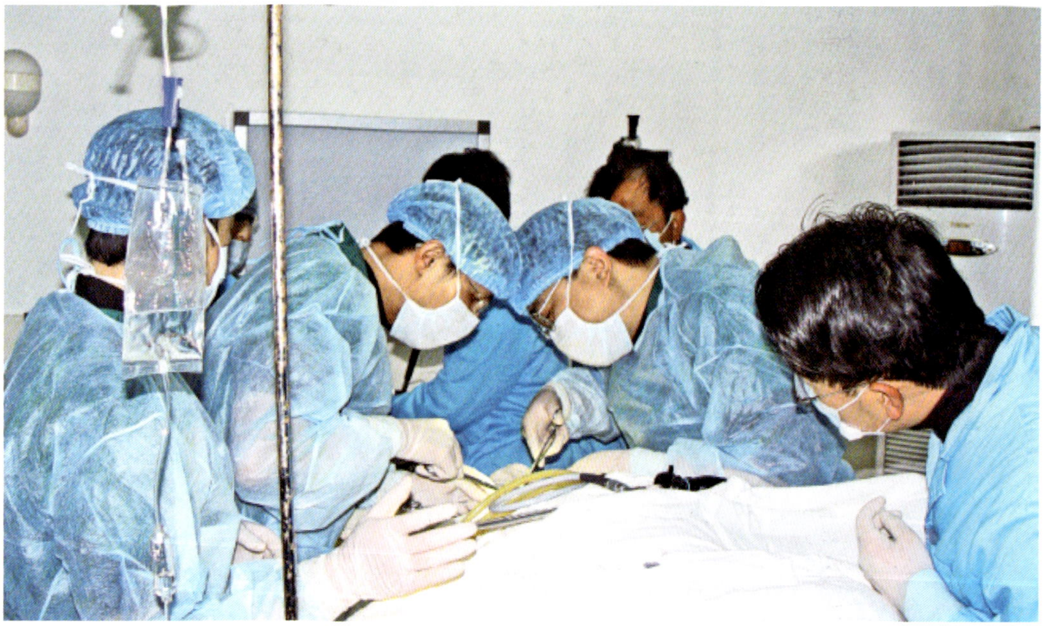

对大熊猫“锦竹”进行腹腔检查

引自四川省档案局、四川省档案学校编《档案见证·大熊猫》，四川人民出版社，2018 年

The giant panda Jin Zhu is receiving an abdominal examination.

Cited from *Archives Witness: The Giant Panda* compiled by Sichuan Province Archives Bureau and Sichuan Archives College, and published by Sichuan People's Publishing House, 2018

口腔疾病

大熊猫口腔炎发病原因比较多样。圈养成年大熊猫多因口腔外伤、齿龈炎或口腔不洁、细菌感染而患病，幼崽则多因奶具、口腔不洁、细菌及真菌等感染所致。其症状多为厌食、拒食、流涎等。

Oral Disease

The giant panda suffers from stomatitis for various reasons, captive adults often from oral trauma, gingivitis, unclean oral cavity and bacterial infection, while cubs are likely to be infected by contaminated milk containers, unclean oral cavity, bacteria and fungi. When infected, they show such symptoms as anorexia, apastia and salivation.

科研人员治疗患胃胀气疾病的大熊猫

引自四川省档案局、四川省档案学校编《档案见证・大熊猫》，四川人民出版社，2018 年

A scientist is treating a giant panda suffering from gastric distention.

Cited from *Archives Witness: The Giant Panda* compiled by Sichuan Province Archives Bureau and Sichuan Archives College, and published by Sichuan People's Publishing House, 2018

胃肠炎

大熊猫胃肠炎分为急性胃肠炎、慢性胃肠炎、出血性胃肠炎等，是大熊猫的常见疾病。圈养大熊猫胃肠炎的发病原因很多，如饲料不清洁、饲料变质、饲料突然改变、精粗饲料配比不当、蛔虫寄生过多等。

Gastroenteritis

Gastroenteritis, subdivided into acute gastroenteritis, chronic gastroenteritis and hemorrhagic gastroenteritis, is a common disease for the giant panda. Various factors may lead to gastroenteritis in captive giant pandas, such as unclean fodder, spoiled fodder, sudden change of fodder, improper proportion of refined and coarse fodder, and too many parasitic roundworms, etc.

谁更美？（2008 年） 张志和 摄

Who looks the prettiest? (2008) Photo by Zhang Zhihe

第五章 调查

在进行有针对性的保护行动之前，有必要对大熊猫及其他珍稀动物进行全面调查摸底。从 20 世纪 70 年代至今，中国共开展了四次大熊猫调查。每次耗时 3 年左右，为中国和全世界的大熊猫保护研究提供了珍贵数据。

全国第一次大熊猫调查从 1974 年 6 月开始，到 1977 年 10 月结束。调查区域包括中国四川、陕西和甘肃三省，这是中华人民共和国成立以来第一次系统调查大熊猫等珍贵动物资源的分布与消长情况。

全国第二次大熊猫调查，由中国和世界自然基金会联合组织，从 1985 年 10 月开始，到 1988 年 11 月结束。与第一次调查数据相比，大熊猫数量明显减少。除了统计误差之外，专家们通过调查数据分析认为，大熊猫栖息地缩小是造成数量下降的最主要原因。

全国第三次大熊猫调查开展时间为 1999—2003 年。这次调查更为全面地掌握了大熊猫分布范围、种群数量等动态变化情况，为此后 10 年的大熊猫保护工作提供了重要参考。

全国第四次大熊猫调查开展时间为 2011—2014 年。2015 年 2 月，国家林业局公布了第四次大熊猫调查结果，掌握了最新的野外种群数量和生存状况。

这四次大熊猫调查为保护大熊猫提供了翔实的资料，也为大熊猫研究提供了真实可靠的依据。

纸本设色《吉祥双宝》, 69cm×138cm（2018 年） 林机 作品

Ink and color on xuan paper: *Lucky Treasures*, 69cm×138cm (2018). By Lin Ji

国画《和谐》, 138cm×69cm（2019 年） 钟远兴 作品

Chinese painting: *Harmony*,138cm×69cm (2019). By Zhong Yuanxing

Chapter 5 Censuses

A thorough census into the giant panda and other rare animals is necessary before any targeted protective action can be taken. China has conducted four national giant panda censuses since the 1970s, each lasting about three years. These censuses have provided valuable data for giant panda conservation.

The first one lasted from June 1974 to October 1977, covering Sichuan, Shaanxi and Gansu. It was the first systematical survey of the distribution and population of the giant panda and other precious animals after the founding of the People's Republic of China in 1949.

The second one, jointly organized by China and WWF, began in October 1985 and ended in November 1988. Compared with the first one, the second saw a sharp decrease in the number of the giant panda. Though there may have been statistical errors, experts believed that it was the shrinking habitat that had led to it.

The third one was conducted from 1999 to 2003. It resulted in a more comprehensive knowledge of the dynamics in the panda's distribution and population, providing an important reference for its protection in the decade to follow.

The fourth one was done from 2011 to 2014. In February 2015, the China's National Forestry and Grassland Administration released the results of the fourth census, providing the latest information about the wild population and its survival status.

The four censuses have provided detailed data and a reliable basis for the research of the giant panda.

大熊猫种群

种群指在一定时间内占据一定空间的同种生物的所有个体。它是进化的基本单位，同一种群的所有生物共用一个基因库。对种群的研究主要是其数量变化与种内关系。按大熊猫分布的六大山系，可将大熊猫栖息地划分为六大部分，与此相应，野生大熊猫可划分成 6 个种群。

Giant Panda Population

Population refers to each and all the same species occupying a given space at a given period. It is the basic unit of evolution. All the population share the same gene pool. The study of population focuses on the quantitative change and intraspecific relationship. The six mountain ranges where the giant panda lives constitute six habitats, hence six populations.

种群密度

大熊猫平均密度表明大熊猫栖息地占有区内单位面积拥有的大熊猫数量（综合统计数）。全国第四次大熊猫调查结果显示，全国大熊猫栖息地面积为 2.58 万平方千米，共生活有1864只野生大熊猫，平均密度为每平方千米0.072只。

Population Density

The average density of giant panda population refers to the number of individuals per unit area in its habitat (synthetic statistics). China's Fourth National Census of shows that 1,864 wild giant pandas live in an area of 25,800km², with an average density being 0.072 individuals per square kilometre.

八宝图（2008 年） 周孟棋 摄

Eight babies (2008). Photo by Zhou Mengqi

从“濒危”到“易危”

2016 年 9 月 4 日，世界自然保护联盟（IUCN）在美国夏威夷宣布将大熊猫的受威胁程度从“濒危”降为“易危”。在此之前，IUCN 濒危物种红色名录中一直将大熊猫划分在“濒危”等级。但是根据中国第四次大熊猫调查数据，IUCN 熊类专家组评估专家认为大熊猫的野生种群在增长，栖息地在恢复，所以决定将它在名录中“降级”。这次“降级”充分肯定了中国政府保护大熊猫所取得的成效。中国为世界生物多样性保护提供了经验。

“降级”并不意味着人类减少了对大熊猫保护的关注度，而是预示着大熊猫这一可爱的物种有着一个光明的未来。

From “Endangered” to “Vulnerable”

On September 4, 2016, the International Union for Conservation of Nature (IUCN) announced in Hawaii, USA, that the giant panda was downgraded form “Endangered” to “Vulnerable” status. Previously, the IUCN’s Red List of Threatened Species had classified the giant panda as an “endangered” species. But according to China’s Fourth National Census of the Giant Panda, the IUCN experts decided to “downgrade” it because of its gradual increase in the wild population and the recovery of its habitat. This is a token of acknowledgement of the efforts and achievements by the Chinese Government in giant panda conservation. China provides the world with practical references in biodiversity protection.

“Downgrading” doesn’t mean less attention to its conservation but a more promising future for this species.

角斗士（2008 年） 周孟棋 摄

Gladiator (2008). Photo by Zhou Mengqi

第六章 科研机构

1978 年，卧龙自然保护区管理局与南充师范学院（今西华师范大学）合作，在保护区臭水沟海拔 2520 米的山上建立了五一棚大熊猫野外生态观察站。1979 年，为救护野生大熊猫，经林业部批准，卧龙自然保护区又在英雄沟海拔 2500 米的山上建立了大熊猫饲养场。

以五一棚大熊猫野外生态观察站和英雄沟大熊猫饲养场为基础，中国政府与世界野生动植物基金会在 1980 年签署协议，成立中国保护大熊猫研究中心，该中心于 1983 年建成并投入使用。

2008 年“5 • 12”汶川特大地震中，中国保护大熊猫研究中心卧龙基地遭到重创。2015 年 12 月 28 日，重新规划的中国大熊猫保护研究中心在四川卧龙国家级自然保护区内正式挂牌成立。研究中心致力于大熊猫的保护和研究工作，并在大熊猫繁育、科研、国际交流合作、大熊猫野化放归等相关领域取得了丰硕成果。

另外一个为大熊猫保护和研究做出突出贡献的科研机构是成都大熊猫繁育研究基地。1983 年开始，邛崃山山系的冷箭竹大面积开花，一部分饥饿的大熊猫被送到成都动物园进行救治，成都大熊猫繁育研究基地在此基础上于 1987 年成立。

除大熊猫繁育、科研、野化放归等研究，成都大熊猫繁育研究基地还在大熊猫保护教育以及大熊猫文化传播领域做出了突出贡献，成为集科研繁育、公众教育和生态旅游于一体的保护研究机构及最佳旅游目的地。

高等院校和科研机构一直是科研的前沿阵地，在大熊猫保护研究领域也不例外。西华师范大学、中国科学院动物研究所、北京大学、四川大学、四川农业大学、四川师范大学、绵阳师范学院、浙江大学等都为大熊猫的保护研究工作贡献了重要力量，取得了大量极具价值的科研成果。

在雪地中穿过障碍物的大熊猫（2016 年）
周孟棋 摄

2018 年 1 月 22 日，大熊猫小种群保护与复壮究开放实验室在四川省雅安市石棉县举行揭仪式 石棉县地方志工作办公室 供图

On January 22, 2018, the Open Laboratory of Giant Panda Small Population Conservation and Rejuvenation Research was launched in Shimiar County, Ya'an City, Sichuan Province. Photo courtesy of Shimian County Local Records Offi

Chapter 6 Research Institutes

Circumventing the obstacle in the snow (2016). Photo by Zhou Mengqi

In 1978, Wolong Nature Reserve Administration and Nanchong Teachers' College (now known as China West Normal University) jointly established Wuyipeng Observation Station at an altitude of 2,520m on Mt. Choushuigou. In 1979, to protect the wild giant panda, the Ministry of Forestry approved the establishment of a breeding farm of the giant panda by Wolong Nature Reserve Administration and it is located 2,500m above sea level on Mt. Yingxionggou.

Based on Wuyipeng Observation Station and the breeding farm on Mt. Yingxionggou, the Chinese Government and WWF agreed, in 1980, to co-fund the project of China Conservation and Research Center for the Giant Panda (China Panda Center, for short), which was completed and came into operation in 1983.

Wolong Base of the China Panda Center was severely damaged by the May 12 Earthquake in Wenchuan in 2008. On December 28, 2015, the newly designed and built China Panda Center officially came into operation in Wolong National Nature Reserve in Sichuan. Committed to the conservation and research of the giant panda, the Center has made remarkable achievements in giant panda breeding, international exchanges, cooperation and reintroduction.

Another research institute that has made outstanding contributions to the giant panda's conservation and research is Chengdu Research Base of Giant Panda Breeding (Chengdu Panda Base, for short). In 1983, extensive flowering of *Bashania fangiana* (a variety of arrow bamboo) in Qionglai Mountains led to starvation and even death of some of the starving ones were sent to Chengdu Zoo. Then in 1987, Chengdu Research Base was founded as required.

In addition to the research of the giant panda breeding and its reintroduction, Chengdu Panda Base has also made great efforts in panda protection education and panda cultural diffusion. And it is now a conservation and research institute and a top tourist destination, integrating scientific breeding, public education and ecological tourism.

Universities and research institutes have always been at the forefront of scientific innovation, and this is also true in the giant panda conservation. China West Normal University, Institute of Zoology of CAS, Peking University, Sichuan University, Sichuan Agricultural University, Sichuan Normal University, Mianyang Teachers' College and Zhejiang University–all of them have made a wealth of valuable achievements in the giant panda conservation.

中国大熊猫保护研究中心都江堰基地（2018 年） 邱宇 摄

Dujiangyan Base of China Conservation and Research Center for the Giant Panda (2018). Photo by Qiu Yu

中国大熊猫保护研究中心

1983 年，中国保护大熊猫研究中心正式建成并投入使用。2015 年，在中国保护大熊猫研究中心基础上成立中国大熊猫保护研究中心，为中国国家林业和草原局直属单位。截至 2019 年，研究中心已拥有卧龙、都江堰和雅安 3 个基地。研究中心在饲养繁殖、疾病控防、野外放归等多个领域取得了重大的研究成果，为促进大熊猫保护事业发展和开展国内外科研合作与文化交流做出巨大贡献。

China Conservation and Research Center for the Giant Panda

It was founded and put into use in 1983. In 2015, it was placed under China's National Forestry and Grassland Administration. Up to 2019, it has branched into three bases in Wolong, Dujiangyan and Ya'an. This center has made great achievements in such fields as captive breeding, disease control and prevention, and reintroduction, contributing greatly to the giant panda conservation, and scientific research cooperation and cultural exchanges between China and other countries.

位于成都市北郊的成都大熊猫繁育研究基地（2007 年） 谭楷 供图

Chengdu Research Base of Giant Panda Breeding, located in the northern suburb of Chengdu (2007). Photo courtesy of Tan Kai

成都大熊猫繁育研究基地

1987 年，成都市建立成都大熊猫繁育研究基地。该机构通过一系列创新性科研工作，攻克了圈养大熊猫人工饲养与管理、繁殖与育幼、疾病防控与种群遗传管理等关键技术难题。2010 年 5 月，成都大熊猫繁育研究基地成立都江堰繁育野放研究中心——“熊猫谷”，启动了大熊猫野外放归研究工作。在大力开展大熊猫科研工作的同时，成都大熊猫繁育研究基地还充分利用丰富的科普教育资源和人才优势开展大熊猫保护教育。

Chengdu Research Base of Giant Panda Breeding

Chengdu Research Base of Giant Panda Breeding (Chengdu Panda Base, for short) was set up in 1987 in Chengdu. It has developed key technologies in artificial feeding and management of captive giant pandas, breeding and cub feeding, disease control and population genetic management, etc. In May 2010, Dujiangyan Base of Giant Panda Breeding and Reintroduction under Chengdu Research Base—Giant Panda Valley—was established, commencing the research on the giant panda's reintroduction. Chengdu Panda Base carries out not only scientific research on the giant panda, but also but also its conservation education by means of educational resources and talent advantages.

西華師范大学 China West Normal University

中国科学院动物研究所 INSTITUTE OF ZOOLOGY,CHINESE ACADEMY OF SCIENCES

西华师范大学

西华师范大学大熊猫研究肇始于20世纪70年代，迄今已历时40余年，是中国大熊猫生态与保护研究的开拓者与奠基者。该校在不同时期分别发表了一系列重要研究成果，为推动大熊猫科学发展、促进大熊猫科学保护做出了杰出贡献，并为中国培养了一大批优秀的保护生物学人才，因此，被誉为“中国的大熊猫大学”。目前，该校以大熊猫种群与栖息地科学保护为出发点，形成了大熊猫保护生态学、大熊猫放归生物学、大熊猫疾病和营养生态学及大熊猫生态系统生态学等研究方向。

China West Normal University

This University commenced its study of the giant panda in the 1970s as the pioneer and founder. With its important research publications over the past forty years, it has made significant contributions to the development of panda science and scientific protection; it has cultivated plenty of outstanding conservation biologists for China, hence the name of “China’s Giant Panda University”. To protect the giant panda population and habitat, the University has been oriented to the research of giant panda conservation ecology, reintroduction biology, pathological and nutrition ecology, and ecosystem ecology.

中国科学院动物研究所

中国科学院动物研究所是以动物科学基础研究为主的社会公益型国家级科研机构。以野生动物和模式动物为研究对象，开展现代动物学研究，服务于人口健康、农业和生物多样性保护等国家重大需求。

中国科学院动物研究所魏辅文院士率领研究团队，以大熊猫为研究模型，取得一系列重大成果：如重建了大熊猫的种群历史，发现其现生种群仍保持着较高的遗传多样性和长期持续的进化潜力；阐明了大熊猫孤立小种群崩溃的生态与遗传机制；推动了中国大熊猫放归和栖息地廊道工程建设。

Institute of Zoology, CAS

Institute of Zoology, Chinese Academy of Sciences, is a state-level non-profit research institution focusing on basic research of animal science. Taking wild animals and model animals as research objects, this organization engages in modern zoological research to serve the major needs of China, such as population health, agriculture and biodiversity protection.

Wei Fuwen, academician with Institute of Zoology, CAS, together with his research team, has made a series of significant breakthroughs on the giant panda research. The team has reconstructed panda’s population history and found that it still maintains a high genetic diversity and sustainable evolutionary potential. And it has elucidated the ecological and genetic mechanisms on how some small isolated populations went bankrupt. This institute also promotes the giant panda’s reintroduction and the construction of habitat corridors in China.

北京大学

1991 年，北京大学大熊猫和野生动物保护研究中心成立，研究方向主要集中在保护生物学领域。经多年研究，北京大学生命与科学学院潘文石教授及其研究团队取得重要科研成果，对大熊猫保护产生了重要作用。

Peking University

Founded in 1991, Giant Panda and Wildlife Conservation Research Center of Peking University focuses on conservation biology. After years of hard work, Professor Pan Wenshi from its School of Life Sciences, together with his research team, has made many significant achievements, which play an important role in the giant panda conservation.

四川大学

在四川大学生命科学学院冯文和教授、冉江洪教授等学者带领下，四川大学在大熊猫生理疾病、生殖遗传、保护遗传学、大熊猫栖息地保护、大熊猫主食竹研究等方面做出突出贡献。

Sichuan University

Feng Wenhe and Ran Jianghong, professors from the College of Life Sciences of Sichuan University, have made extraordinary contributions to the study of the giant panda's physiological disease, reproduction and genetics, conservation genetics, habitat, and staple edible bamboo.

大熊猫“奥莉奥”在木架上玩耍（2017 年） 黄馨 摄
The giant panda Oreo is having fun on the log (2017). Photo by Huang Xin

中国大熊猫研究的标志性人物，西华师范大学生命科学学院胡锦矗教授在大熊猫实验室对大熊猫头骨化石进行研究 引自周孟棋摄影集《中国大熊猫》，四川美术出版社，2009 年

Hu Jinchu, professor of College of Life Sciences at China West Normal University and iconic figure in China's giant panda research, is studying giant panda skull fossils in the giant panda laboratory. Cited from Zhou Mengqi, *China's Giant Panda*, Sichuan Fine Arts Publishing House, 2009

中国科学院院士魏辅文研究员在观察野生大熊猫进食 引自魏辅文著《野生大熊猫科学探秘》，科学出版社，2018 年

Wei Fuwen, academician of Chinese Academy of Sciences, is observing a wild giant panda eating bamboo. Cited from Wei Fuwen, *Scientific Exploration of Wild Giant Pandas*, Science Press, 2018

大熊猫研究者

大熊猫数量稳中有升，这离不开科研工作者们持续不断的努力。以胡锦矗、夏勒、秦自生、朱靖、潘文石、张安居、冯文和、刘维新、魏辅文、张和民、张志和、冉江洪、张泽钧等为代表的中外研究者们，在诸多研究领域取得了突破性进展，并推动了中国大熊猫保护事业的发展。

Panda Researchers

The persistent efforts by panda researchers have resulted in a steady increase in its number. Chinese and foreign researchers, such as Hu Jinchu, George B. Schaller, Qin Zisheng, Zhu Jing, Pan Wenshi, Zhang Anju, Feng Wenhe, Liu Weixin, Wei Fuwen, Zhang Hemin, Zhang Zhihe, Ran Jianghong and Zhang Zejun have made breakthroughs in this field and furthered China's giant panda conservation.

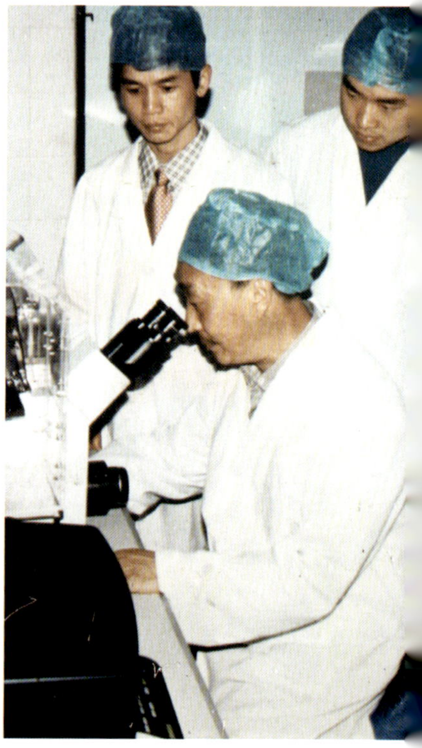

1	2	3
4	5	

1. 胡锦矗（右）与魏辅文（左）在大熊猫科学发现地——宝兴县邓池沟天主教堂留影 引自魏辅文著《野生大熊猫科学探秘》，科学出版社，2018 年

2. 西华师范大学秦自生教授近距离观察大熊猫活动 秦自生 供图

3. 中国科学院古脊椎动物与古人类研究所黄万波教授在禄丰始熊猫化石地层采集资料 引自谭楷著《熊猫的故事》，五洲传播出版社，2015 年

4. 北京大学生命科学学院教授潘文石（左），美国动物学家、博物学家乔治·夏勒（右），北京大学保护生物学教授吕植（中）近距离观察大熊猫活动 引自谭楷著《熊猫的故事》，五洲传播出版社，2015 年

5. 2007 年，中国大熊猫繁育技术委员会主任张安居教授带领年轻人开展科研攻关 成都大熊猫繁育研究基地 供图

1. Hu Jinchu (Right) and Wei Fuwen (Left) at Dengchigou Catholic Church. Cited from Wei Fuwen, *Scientific Exploration of Wild Giant Pandas*, Science Press, 2018

2. Professor Qin Zisheng is making close observations. Photo courtesy of Qin Zisheng

3. Professor Huang Wanbo from the Institute of Vertebrate Paleontology and Paleoanthropology of CAS is collecting fossils from the fossil strata of *Ailurarctos lufengensis*. Cited from Tan Kai, *Story of the Giant Panda*, China Intercontinental Press, 2015

4. Professor Pan Wenshi from School of Life Science, Peking University (left), George B. Schaller, American biologist and naturalist (right), Professor Lv Zhi, conservation biologist from Peking University (centre), are observing giant panda's activities at close range. Cited from Tan Kai, *Story of the Giant Panda*, China Intercontinental Press, 2015

5. Professor Zhang Anju was leading young people to crack technical problems in 2007. Photo courtesy of Chengdu Research Base of Giant Panda Breeding

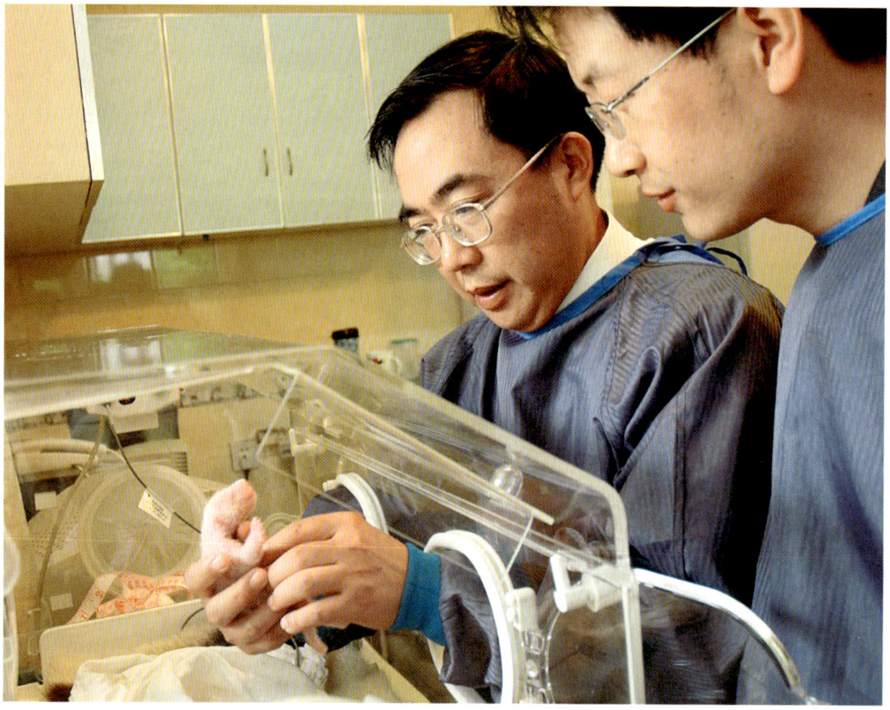

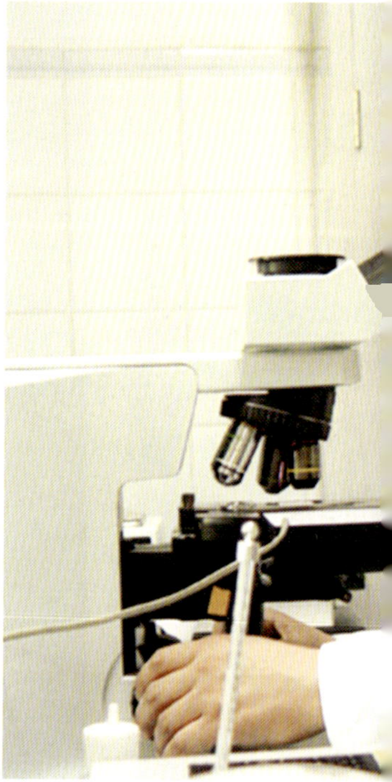

1	2	3	
4	5	6	7

1. 2001 年，中国保护大熊猫研究中心主任张和民（左）正在检查人工育幼小崽的发育状况 卧龙核桃坪基地 供图

2. 2017 年，成都大熊猫繁育研究基地主任张志和正在进行科学研究 成都大熊猫繁育研究基地 供图

3. 1992 年，中国首次对大熊猫群体进行全面体检。图为成都军区总医院主任医师曾祥元（中）在观察大熊猫眼底及微循环情况 新华社 供图

4. 2002 年 10 月 17 日，四川大学生命科学学院冉江洪教授带队在安县千佛山国家级自然保护区进行大熊猫野外调查 冉江洪 供图

5. 2012 年，西华师范大学珍稀动植物研究所、生命科学学院研究员张泽钧在甘肃省白水江国家级自然保护区进行野外考察 张泽钧 供图

6. 20 世纪 90 年代，成都大熊猫繁育研究基地高级兽医师叶志勇（左）和他的助手用冷冻精液为大熊猫进行人工授精 新华社 供图

7. 1994 年，四川大学生命科学学院冯文和教授在成都大熊猫繁育研究基地对大熊猫进行生理测定 四川大学图书馆 供图

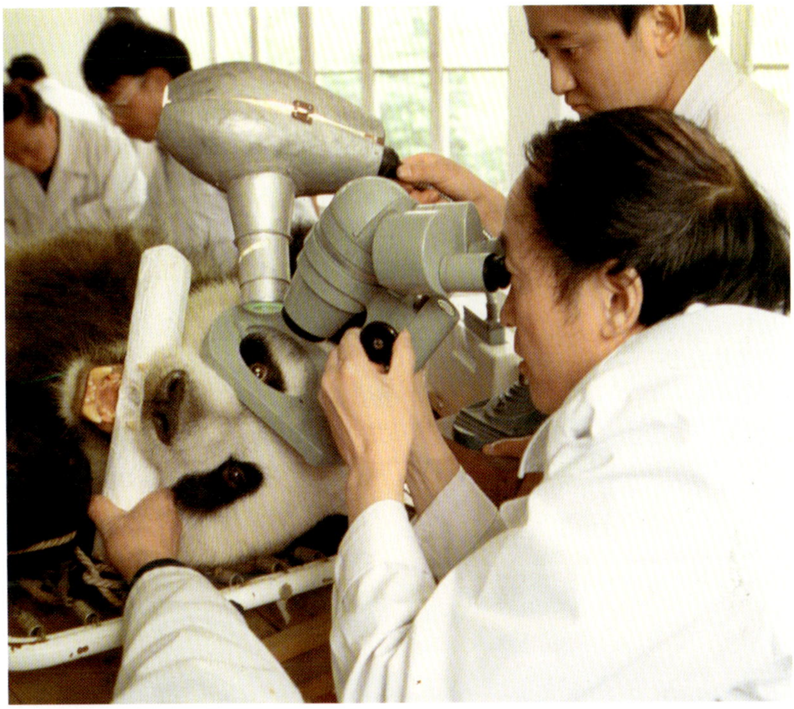

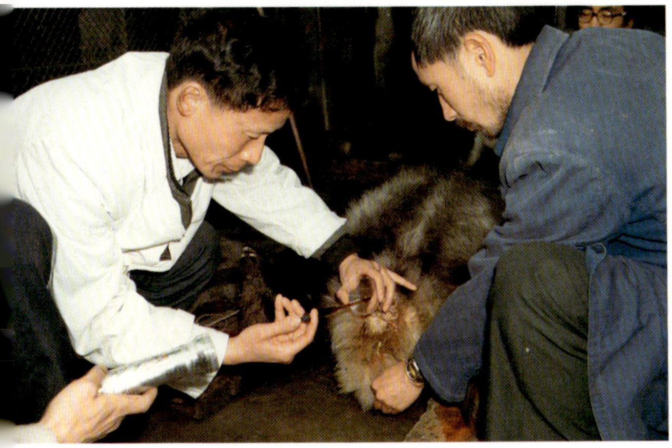

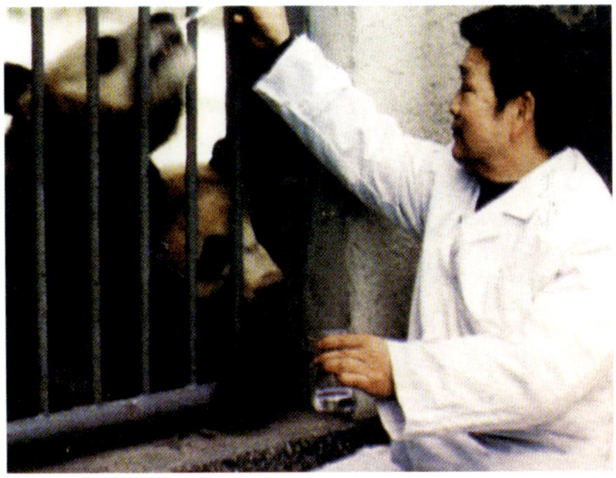

1.Zhang Hemin (left), Executive Vice Director of China Conservation and Research Center of the Giant Panda, was examining an artificially bred cub in 2001. Photo courtesy of Hetaoping Base at Wolong

2.Zhang Zhihe, Director of Chengdu Research Base of Giant Panda Breeding, was conducting scientific research in 2017. Photo courtesy of Chengdu Research Base of Giant Panda Breeding

3.In 1992, for the first time in China, a thorough physical examination was carried out for the giant panda population. Zeng Xiangyuan (middle), archiater of Chengdu Junqu General Hospital, was examining the fundus and microcirculation of a giant panda. Photo courtesy of Xinhua News Agency

4.On October 17, 2002, Professor Ran Jianghong of College of Life Science, Sichuan University, was conducting a field survey of giant pandas in Qianfoshan National Nature Reserve, Anxian County. Photo courtesy of Ran Jianghong

5.Zhang Zejun, researcher of the Institute of Rare Animals and Botany and College of Life Science of China West Normal University, was on a field trip in Baishuijiang National Nature Reserve, Gansu Province in 2012. Photo courtesy of Zhang Zejun

6.In the 1990s, Ye Zhiyong (left), senior veterinarian at Chengdu Research Base of Giant Panda Breeding, were his assistant, are using frozen semen for insemination. Photo courtesy of Xinhua News Agency

7.Professor Feng Wenhe was conducting a physical examination of a giant panda at Chengdu Research Base of Giant Panda Breeding in 1994. Photo courtesy of Sichuan University Library

1981 年，中外专家在五一棚讨论考察计划 周守德 摄

In 1981, Chinese and foreign experts at Wuyipeng Observation Station were discussing the investigation plan. Photo by Zhou Shoude

中外研究合作

中外科学家为研究和保护大熊猫开展了大量合作，并开展了多层次的学术交流。在 1993 年举行的成都国际大熊猫保护学术研讨会，有近百名中外专家参与。近年来中外交流更加频繁，并且从学术交流扩展到学术、文化等多方面的交流活动。

Research Cooperation with Other Countries

Scientists from China and other countries have cooperated in the research and conservation of the giant panda and carried out academic exchanges at multiple levels. In 1993, around 100 Chinese and foreign experts attended Chengdu International Symposium of Giant Panda Conservation. In recent years, more such exchanges, ranging from scholarship to culture, have been conducted between them.

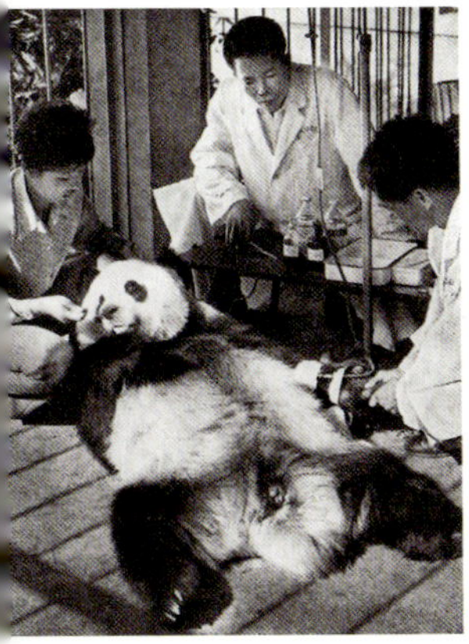

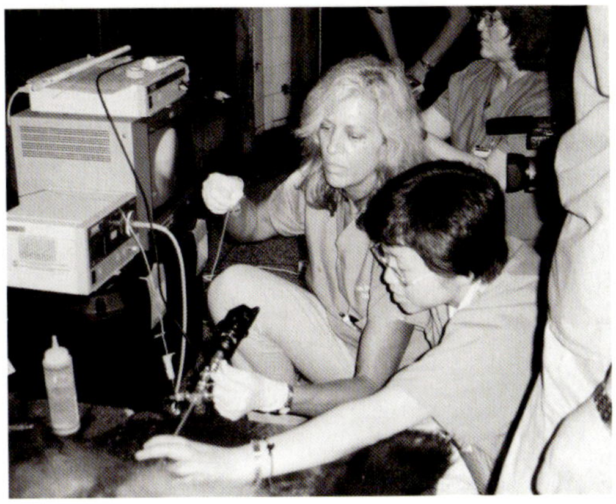

1		2
3	4	5

1. 1993 年，在成都召开成都国际大熊猫保护学术研讨会 引自成都动物园、成都大熊猫繁育研究基地编《成都国际大熊猫保护学术研讨会论文集》，四川科学技术出版社，1994 年

2. 1985 年 11 月，秦自生教授（左）与美国植物学家艾伦•泰勒（右）鉴别箭竹种类 秦自生 供图

3. 1981 年，福州动物园开展大熊猫静脉输液演习 常素琴 摄

4. 中外合作研究大熊猫（2005 年）谭楷 供图

5. 1980 年，世界野生动植物基金会小组的代表斯科特爵士（左）、夏勒博士（右）在卧龙自然保护区观察大熊猫粪便 新华社 供图

1.Chengdu International Symposium on Giant Panda Conservation in 1993. Cited from *Proceedings of Chengdu International Symposium on Giant Panda Protection* compiled by Chengdu Research Base of Giant Panda Breeding in 1994

2.In November 1985, Qin Zisheng (left) and American botanist Alan Taylor (right) were identifying arrow bamboo species. Photo courtesy of Qin Zisheng

3.Giant Panda venous transfusion Exercise was going on in Fuzhou Zoo in 1981. Photo by Chang Suqin

4.Sino-foreign cooperation on the giant panda conservation program (2005). Photo courtesy of Tan Kai

5.In 1980, Sir Scott (left) and Dr. Schaller (right), representatives of the WWF, were examining giant panda feces in Wolong Nature Reserve. Photo courtesy of Xinhua News Agency

第五篇

保护

青神竹编《远眺眺》，直径 40cm（2009 年） 陈云华 作品

Bamboo weaving: *Higher and Farther*, 40 cm in diameter (2009). By Chen Yunhua

Part 5

Protection

人类的伟大不在于征服，而在于拯救

当我们看见大熊猫的黑眸，

亦如看见人类深邃的内心。

爱是看不见的羁绊，

是世界上强大的力量。

当人类学会了守护，才有可能改变自然。

It is protection rather than conquest that makes mankind a great race

One gaze at the dark eyes of a giant panda and every one of us is stirred inside.

Love, that almighty power on earth,

is an invisible bond between man and other creatures in nature;

this power drives man to the protection and preservation of that species.

第一章 救灾

1975 年，在中国四川省多地出现大熊猫自然死亡现象。林业部门对此十分重视，并迅速组织专家展开调查。据调查统计，1975 年冬至 1976 年春末，全国野生大熊猫自然死亡数量达 92 只，这些大熊猫的死亡大都源于饥饿。

调查发现，当时大熊猫最喜爱的冷箭竹开花枯死面积超过 95%，并且在 1975 年底，四川盆地普遍降温，饥寒交迫的大熊猫没有熬过那个冬天。四川随即展开对灾区存活大熊猫的抢救保护。

无独有偶，从 1983 年 5 月开始，在卧龙国家级自然保护区、宝兴蜂桶寨国家级自然保护区、九寨沟自然保护区等地，接连出现箭竹大面积开花枯死现象，大熊猫又一次面临生存劫难。主食竹开花引发大熊猫生存难的问题受到全世界关注，救援工作随即展开。中国各地自发捐款，同时全球也掀起了抢救大熊猫的热潮。

两次拯救缺食大熊猫只是保护工作的一个缩影。在 2008 年“5·12”汶川特大地震中，拯救被困大熊猫也是保护史上浓墨重彩的一笔。当时地处重灾区的中国保护大熊猫研究中心遭到重创，研究中心围墙倒塌，63 只圈养大熊猫被困在废墟之中，所幸经抢救脱险。

抢救性保护只能解决燃眉之急，更重要的是恢复生态，保护大熊猫栖息地。中国积极建立自然保护区、大熊猫国家公园，希望从根本上保护大熊猫的生存环境。

保护大熊猫不仅是在拯救这一生灵，也是在唤醒人们“保护生态环境，珍爱野生动物”的意识。人们携手保护的不仅是大熊猫，更是包括人类在内的整个自然界。

上 20 世纪 90 年代初期，宝兴县蜂桶寨国级自然保护区工作人员在野外发现病饿大猫，并积极开展救治，图为组织运救途中水的场景 高华康 摄

下 “5·12”汶川特大地震发生后，中国护大熊猫研究中心的科研人员冒着生命危奋力抢救大熊猫 引自四川省档案局、四川档案学校编《档案见证·大熊猫》，四川民出版社，2018 年

Chapter 1 Rescue

per) In the early 1990s, the staff of Fengtongzhai National Nature Reserve, Baoxing County found a sick and hungry giant panda and undertook its rescue. The picture shows the dangerous transfer process. Photo by Gao Huakang

(Below) Staff members of China Conservation and esearch Center for the Giant Panda risk their lives to save giant pandas after a massive earthquake struck Wenchuan. Cited from *Archives Witness: The Giant da* compiled by Sichuan Province Archives Bureau and Sichuan Archives College, and published by Sichuan People's Publishing House in 2018

In 1975 deaths of some giant pandas were reported in many places of Sichuan. China's National Forestry and Grassland Administration responded quickly by organizing experts to investigate why. They found that ninety-two wild giant pandas died from lack of food sources from the winter of 1975 to the late spring of 1976.

The experts also found that 95% of *Bashania fangiana*, the giant panda's favorite fodder bamboo, had flowered and died and few giant pandas survived the drop in temperature at the end of that year in Sichuan Basin. Subsequently, rescue of suffering giant pandas got into full swing.

Starting from May 1983, *fargesia spathacea* (arrow bamboo) died after flowering in Wolong National Nature Reserve, Fengtongzhai National Nature Reserve, Jiuzhaigou National Nature Reserve and others. The giant panda was faced with another challenge of survival. This grabbed the world's attention, followed by rescue efforts. Donations were made across China, and a surge of enthusiasm for saving the giant panda began to prevail globally.

These efforts were only part of the story. Another significant endeavor worth remembering was the successful transfer of 63 captive giant pandas trapped in the China Panda Center devastated by the catastrophic May 12, 2008 Wenchuan Earthquake.

However, rescuing is not the best policy; what is more important is to restore its natural habitat. China has established nature reserves and giant panda national parks as its shelters.

Protecting the giant panda does not just mean its security, but a means for heightening the public awareness of "protecting environment, cherishing wildlife." By protecting it, we're helping to safeguard the broader environment, which man and other species depend on.

2010年，在宝兴县五龙乡东升村村民家中啃食农家鲜猪骨头的大熊猫 高华康 摄

A giant panda was nibbling at swine bone at a farmhouse in Dongsheng Village, Wulong Township, Baoxing County in 2010. Photo by Gao Huakang

1975年之灾

1975年前后竹子开花，岷山山系受灾严重。其受灾面积约5250平方千米，占岷山大熊猫分布区的40%。灾区大熊猫数量占6个重点调查县总数的76.5%，相当于秦岭和大凉山大熊猫分布区的总和。

Disaster in 1975

In 1975 bamboo bloomed in Minshan Mountains. The disaster hit an area of about 5,250km², 40% of the panda habitat there. The number of victimized pandas accounted for 76.5% of the six major habitat counties, equaling the total number in Qinling and Daliang Mountains.

2017 年 4 月 19 日，一只野生病饿大熊猫在宝兴县中坝乡紫云村的山林中被发现并得到救助 高华康 摄

On April 19, 2017, a wild sick panda was found and rescued in the forest of Ziyun Village, Zhongba Township, Baoxing County. Photo by Gao Huakang

救护病饿、伤残大熊猫

1983 年 9 月至 1989 年 12 月，宝兴县蜂桶寨、青川县唐家河、平武县王朗、北川县小寨子沟、天全县喇叭河以及卧龙等自然保护区的饲养场及其他地区，共抢救、治疗了 83 只病饿、伤残大熊猫，63 只幸运地活了下来。其中，有 33 只被收养，30 只被放归野外。

Rescuing Sick, Starving and Wounded Giant Pandas

From September 1983 to December 1989, 83 giant pandas, which were sick, starving, wounded or disabled, were cured in the feedlots in such nature reserves as Fengtongzhai, Tangjiahe, Wanglang, Xiaozhaizigou, Lapa River and Wolong. 63 of them survived, 33 of which were kept in captivity and 30 sent back to their habitat upon recovery.

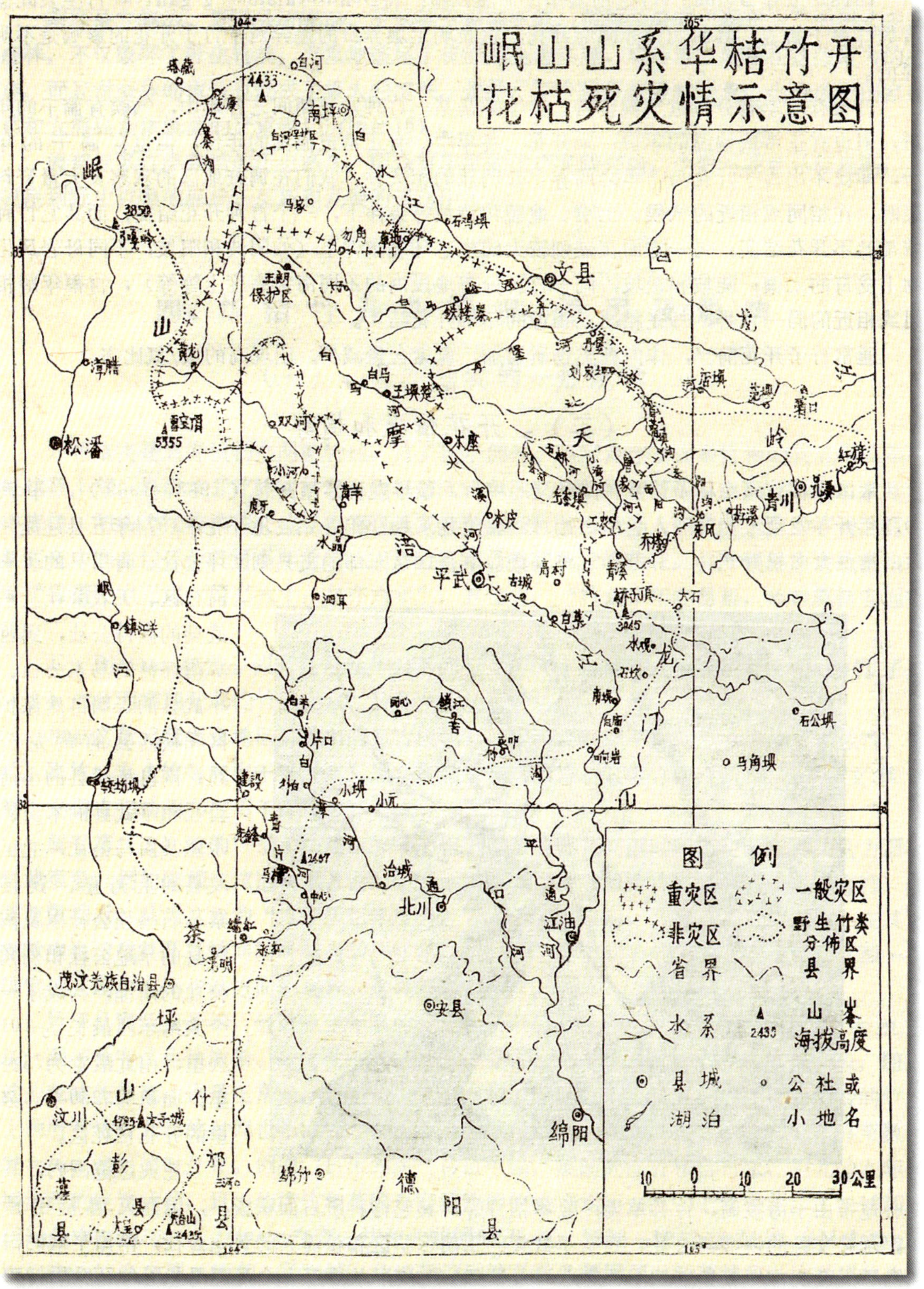

20世纪70年代，王绍庆手绘的岷山山系华桔竹开花枯死灾情示意图　张奕 供图

Wang Shaoqing's sketch of the flowering of *Fargesia spathacea Franeh* in Minshan Mountains in the 1970s. Photo courtesy of Zhang Yi

左 20 世纪 80 年代，胡锦矗在王朗自然保护区开展野外研究工作 绵阳市林业局 供图

右 20 世纪 80 年代，中外科学家在五一棚进行鲜竹称量 孙志江 摄

(Left) In the 1980s, Hu jinchu and other panda experts were conducting field research in the Wanglang National Nature Reserve. Photo courtesy of Mianyang Forestry Administration

(Right) In the 1980s, scientists were measuring fresh bamboos at Wuyipeng Observation Station. Photo by Sun Zhijiang

1983 年竹子开花

1983 年，卧龙国家级自然保护区的工作人员深入大熊猫主要栖息地，发现共有 320 平方千米冷箭竹开花，达自然保护区全部面积的 95%，140 余只大熊猫基本失去食物来源。此外，1983 年 9 月初步统计，雅安地区冷箭竹开花面积达 134 平方千米，青川县 100 平方千米箭竹中，约有 67 平方千米开花枯死。

Bamboo Flowering in 1983

In 1983, the staffers in Wolong National Nature Reserve found while patrolling the major panda habitat that, of the 320km^2 of Bashania fangiana (a variety of arrow bamboo), which accounts for 95% of the nature reserve flowered, and over 140 pandas lost their food sources. According to the statistics in September 1983, up to 134km^2 of Bashania fangiana bloomed in Ya'an. About 67 out of the 100km^2 of arrow bamboo bloomed and died in Qingchuan County.

中国四川省丹巴县部分地区曾为大熊猫栖息地，但在全国第三次和第四次大熊猫调查时，丹巴县未发现大熊猫。图为丹巴县美人谷（2010年）周孟棋 摄

While part of Danba County in Sichuan Province used to a habitat for the giant panda, there was no trace of this species during the third and fourth national censuses. The picture shows the Beauty Valley in Danba County (2010). Photo by Zhou Mengqi

卧龙国家级自然保护区内海拔4200米的高山海子 谭楷 供图

Alpine lakes in Wolong National Nature Reserve at 4,200m above sea level. Photo courtesy of Tan Kai

卧龙国家级自然保护区

地处川西平原与青藏高原的过渡地带，森林资源丰富，珍稀动植物种类繁多。境内有近 1900 种高等植物，109 种兽类，365 种鸟类，1700 多种昆虫，被誉为“天然生物广谱基因库”。1983 年 3 月，卧龙国家级自然保护区在辖区内成立卧龙特别行政区。2006 年，卧龙国家级自然保护区随“四川大熊猫栖息地”被列入世界自然遗产名录。

Wolong National Nature Reserve

Located in the transitional zone between the Chengdu Plain and Qinghai-Tibet Plateau, Wolong National Nature Reserve abounds in forest resources and boasts a wide assortment of rare animal and plant species. It is crowned as a “natural biological gene bank” with nearly 1,900 higher plants, 109 animal species, 365 bird species and over 1,700 kinds of insects. In March 1983, Wolong Special Administrative Region was established under the jurisdiction of Wolong National Nature Reserve. In 2006, Wolong National Nature Reserve was inscribed on the World Heritage List together with “Sichuan Giant Panda Sanctuaries”.

王朗大草坪 王朗国家级自然保护区 供图

Wanglang Grassland. Photo courtesy of Wanglang National Nature Reserve

王朗国家级自然保护区

位于中国四川省绵阳市平武县境内，总面积约 323 平方千米。建于 1963 年，2002 年晋升为国家级自然保护区。该保护区是中国建立最早的 4 个以保护大熊猫等珍稀野生动物及其栖息地为主的自然保护区之一。

Wanglang National Nature Reserve

Located in Pingwu County, Sichuan Province, it occupies an area of about $323km^2$. Established in 1963, it became a national nature reserve in 2002, one of the first four nature reserves focusing on protecting rare species (such as the giant panda) and their habitats.

九寨沟诺日朗瀑布 田捷砚 摄

Nuorilang Waterfall in Jiuzhaigou. Photo by Tian Jieyan

九寨沟国家级自然保护区

位于岷山山脉南段朵尔纳峰北麓，主要景观是色彩如蓝宝石的流水及彩湖叠瀑。沟内有 108 个梯级湖泊，17 处瀑布，分布着 2000 多种高等植物，300 余种野生脊椎动物。其中，大熊猫、川金丝猴、扭角羚等为国家一级保护动物。1992 年 12 月，九寨沟自然保护区被列入世界自然遗产名录。

Jiuzhaigou National Nature Reserve

Located at the northern foot of Duoerna Peak of Minshan Mountains, it boasts sapphire-like water, colorful lakes and spectacular waterfalls. There are 108 ladder-like lakes, 17 magnificent waterfalls, over 2,000 higher plants and about 300 wild vertebrates; giant pandas, golden snub-nosed monkeys, takins and other rare animals have been put under first-class state protection. In December 1992, Jiuzhaigou National Nature Reserve was included on the World Heritage List.

唐家河国家级自然保护区

位于中国四川省广元市青川县境内，有400余平方千米的林区，自然条件优越。保护区内空气湿度很高，有充足的水源，是珍稀动物大熊猫的主要栖息和繁殖地。此外，扭角羚、川金丝猴及珙桐、连香树等珍稀动植物亦在此分布。

Tangjiahe National Nature Reserve

Located in Qingchuan County, Guangyuan City, Sichuan Province, Tangjiahe National Nature Reserve enjoys excellent natural conditions, as evidenced by over 400km^2 of forests. High humidity and adequate water supply make this place a main habitat and breeding area for the giant panda. This nature reserve is also home to other rare animals and plants like takins, golden snub-nosed monkeys, dove trees, and katsure trees.

唐家河国家级自然保护区（2016年） 周孟棋 摄

Tangjiahe National Nature Reserve (2016). Photo by Zhou Mengqi

佛坪国家级自然保护区（2013 年） 周孟棋 摄

Foping National Nature Reserve (2013). Photo by Zhou Mengqi

佛坪国家级自然保护区

位于中国陕西省秦岭中段南坡，是以保护大熊猫为主的森林和野生动物类型的自然保护区，区划面积 350 平方千米。区内有 1400 余种植物，49 种兽类，162 种鸟类，22 种两栖爬行类动物。其中，国家一级保护动物有大熊猫、川金丝猴、扭角羚、金雕、豹等。

Foping National Nature Reserve

Located on the south slope of the middle section of Qinling Mountains in Shaanxi, Foping National Nature Reserve is a forestry and wildlife nature reserve focusing on protecting the giant panda, covering an area of 350km². It is home to about 1,400 species of plant, 49 species of animal, 162 species of bird, and 22 species of amphibian reptile, among which the giant panda, golden snub-nosed monkey, takin, golden eagle and leopard are under first-class state protection.

龙溪—虹口国家级自然保护区（2003 年） 周孟祺 摄
Longxi-Hongkou National Nature Reserve (2003). Photo by Zhou Mengqi

第三章 栖息地

全国第四次大熊猫调查数据显示，中国大熊猫栖息地总面积约 2.58 万平方千米。其中，四川省大熊猫栖息地面积 2.03 万平方千米，约占大熊猫栖息地总面积的 78.7%。

这 2.58 万平方千米的栖息地，因受到森林采伐、农业生产和交通建设等影响，并非整体成片。大熊猫种群随着栖息地的分隔而孤立，大熊猫保护面临严峻考验。恢复栖息地的连通，有助于提高大熊猫种群之间的自然远缘交配频率，避免小种群带来的遗传衰退，所以保持现有栖息地面积不减少，并促进栖息地人工恢复显得尤为重要。

相关部门已经提出“大熊猫走廊带建设”计划。大熊猫走廊带位于两块大熊猫栖息地之间最容易被大熊猫等野生动物利用的地方，需要先形成自然森林覆盖层，还需考虑走廊带人为干扰消除等因素。大熊猫的分布区内共规划有 15 条走廊带，除了分布在中国陕西省太白县境内的沙坝—皂角湾走廊带和分布在陕西省佛坪县境内的龙草坪走廊带之外，其余如小河廊道、二郎山—大相岭廊道群等 13 条走廊带都分布在四川省境内。

除现有栖息地之外，还有潜在栖息地。中国有 0.91 万平方千米的大熊猫潜在栖息地，分布在四川、陕西、甘肃三省。其中，四川省的大熊猫潜在栖息地面积最大，为 0.41 万平方千米，占中国潜在栖息地总面积的 45.05%。经过多年努力，大熊猫栖息地正得到越来越多的保护。

蜀绣《单面绣红石熊猫》
80cm×36cm（2015 年）
杨德全 作品
Sichuan Embroidery: *A Giant Panda on Redstone*,
80cm×36cm (2015).
By Yang Dequan

Chapter 3 Habitat

蜀锦《熊猫春嬉图》
106cm×57cm（2010 年）
胡光俊 作品
Sichuan Figured Satin: *Frolicking Giant Pandas on a Spring Day*,
106cm×57cm (2010).
By Hu Guangjun

According to China's Fourth National Census of the Giant Panda, the giant panda's habitat totals about 25,800km², about 78.7% (20,300km²) in Sichuan.

This habitat, however, is fragmented due to deforestation, agricultural production, road construction and so on. As a result, the giant panda population in one reserve is isolated from that in another, which poses a challenge for its protection. Restoring the connection between the patches of habitat can promote outbreeding and slow down genetic decline of the small population. As such, it is important to prevent the existing habitat from dwindling and restore its connection by human efforts.

The governmental organizations concerned have put forward the plan of "Bamboo Corridors for the Giant Panda," which refer to the stripes of land between fragmented habitats for wildlife including the giant panda. The establishment of such corridors means natural forestry coverage and absence of human disturbance. There are 15 corridors projected in the range of giant pandas, 13 of which are in Sichuan like Xiaohe and Erlangshan-Daxiangling Corridors, while the other two, Shaba-Zaojiaowan Corridor in Taibai County and Longcaoping Corridor in Foping County, are in Shaanxi.

Besides the existing habitat, there are potential ones. The censuses indicate that the potential giant panda habitat covers 9,100km², mainly in Sichuan, Shaanxi and Gansu, the largest in Sichuan, accounting for 45.05% (4,100km²) of the total. Years of effort have put the giant panda habitat under better protection.

岷山（2008 年） 周孟棋 摄

Minshan Mountains (2008). Photo by Zhou Mengqi

献给地球的礼物

“献给地球的礼物”这一奖项是世界自然基金会为展示全球自然保护的成就，于 1996 年设立的。2003 年，世界自然基金会授予陕西省人民政府“献给地球的礼物”荣誉证书，以表彰其在秦岭大熊猫保护工作中的突出贡献。秦岭成为全球第 83 份“献给地球的礼物”，这是中国第 2 份“献给地球的礼物”。

2006 年，世界自然基金会将岷山山系大熊猫栖息地评为“献给地球的礼物”，以表彰四川省人民政府、甘肃省人民政府在保护大熊猫和生物多样性方面做出的突出贡献。这是全球第 103 份，也是中国第 5 份“献给地球的礼物”。

Gift to the Earth

Gift to the Earth (GTTE) award was set up in 1996 by WWF to honor important conservation achievements. In 2003, GTTE went to Shaanxi Provincial Government, as a token of commendation of its special contributions to the giant panda conservation in Qinling. Thereupon, Qingling became the 83th “Gift to the Earth” in the world, and the 2nd in China.

In 2006, GTTE went to the Giant Panda Habitat in Minshan Mountains in recognition of the significant contributions by the governments of Sichuan and Gansu to the giant panda conservation and biodiversity. This is the 103rd gift in the world and the 5th in China.

卧龙西河中段针阔混交林中的大熊猫（2014 年） 杨文刚 摄

A giant panda in coniferous-broadleaved forests in the middle section of Xihe River,Wolong National Nature Reserve(2014). Photo by Yang Wengang

栖息地植被类型

中国大熊猫栖息地的植被可分为两个类型：一是自然植被，分为针叶林、阔叶林、灌丛和草甸 4 大植被型组，13 个植被型，21 个植被亚型；二是栽培植被，分为 3 个植被型、10 个植被亚型。

Types of Habitat Vegetation

The vegetation in the habitat falls into two categories: natural and cultivated vegetations. The first consists of 4 type groups (coniferous forest, broad-leaved forest, bushes and meadows), 13 types and 21 subtypes. The second has 3 types and 10 subtypes.

四川省大熊猫栖息地廊道分布图
Distribution of Giant Panda Habitat Corridors in Sichuan Province

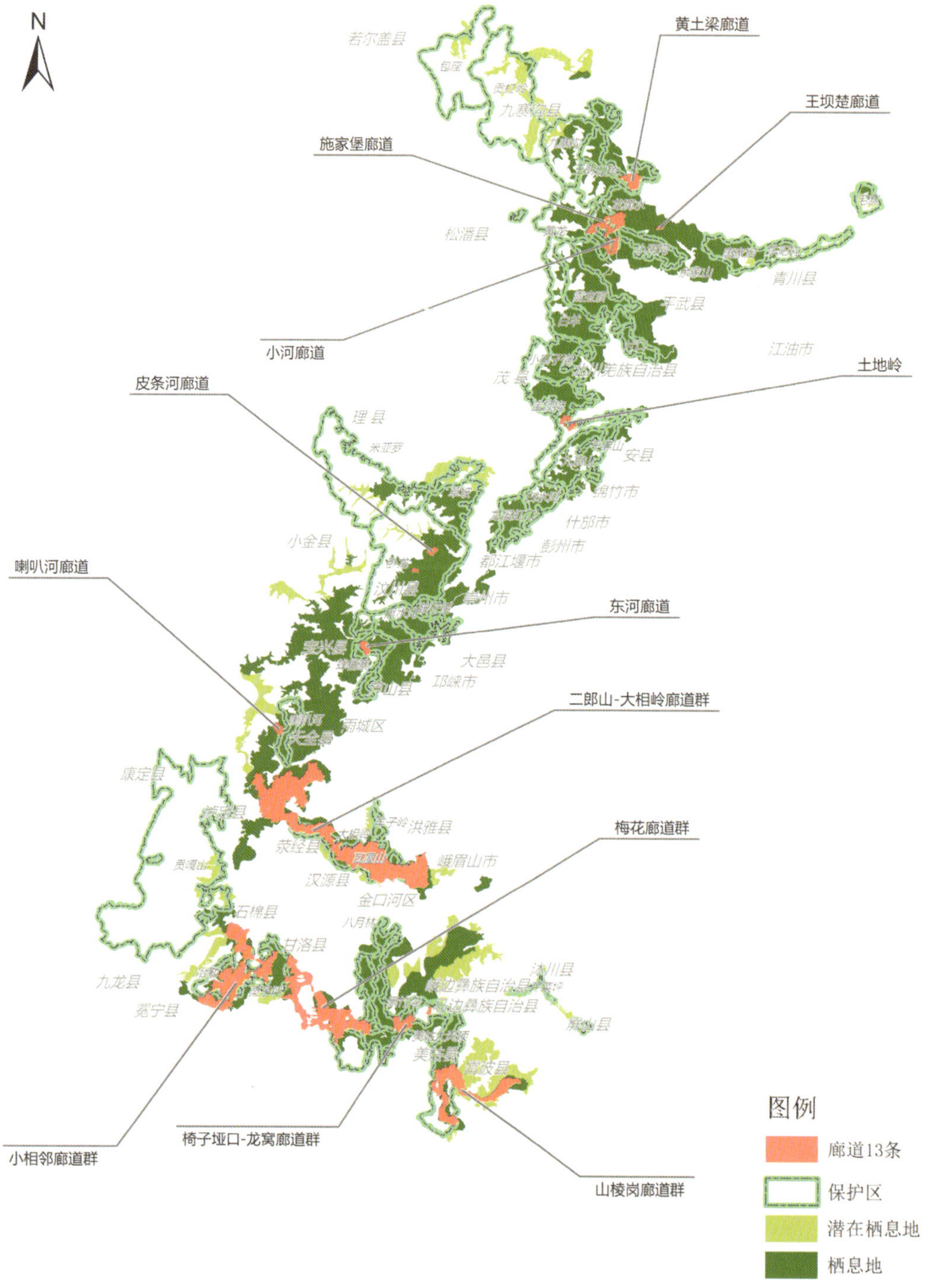

四川省大熊猫栖息地廊道分布图 四川省林业和草原局 供图

Distribution of Giant Panda Habitat Corridors in Sichuan Province. Photo courtesy of Sichuan Province Forestry and Grassland Administration

2006 年 5 月 14 日，世界自然基金会（WWF）中国首席代表欧达梦（右二）与志愿者在秦岭亚种大熊猫绿色走廊建设工程启动仪式上栽种秦岭箭竹 新华社 供图

On May 14, 2006, Dermot O' Gorman (2nd from right) Chief Representative of the World Wide Fund for Nature (WWF) in China, together with volunteers planted Qinling arrow bamboos at the launch ceremony of the green corridor construction project for the Qinling subspecies of the giant panda. Photo courtesy of Xinhua News Agency

大熊猫走廊带（道）

由于人为原因，大熊猫栖息地日益破碎化，这严重威胁了大熊猫的生存。建立大熊猫走廊带（道）能有效连接各个大熊猫栖息地，为大熊猫进行基因交流提供更大的可能。目前，大熊猫的分布区内共规划有 15 条走廊带。

Habitat Corridors for the Giant Panda

Human development has fragmented the giant panda habitat, which poses a serious threat to its survival. The Habitat Corridors for the Giant Pandas can effectively connect isolated patches and increase the possibility of genetic flow. Fifteen corridors have been projected across the habitat.

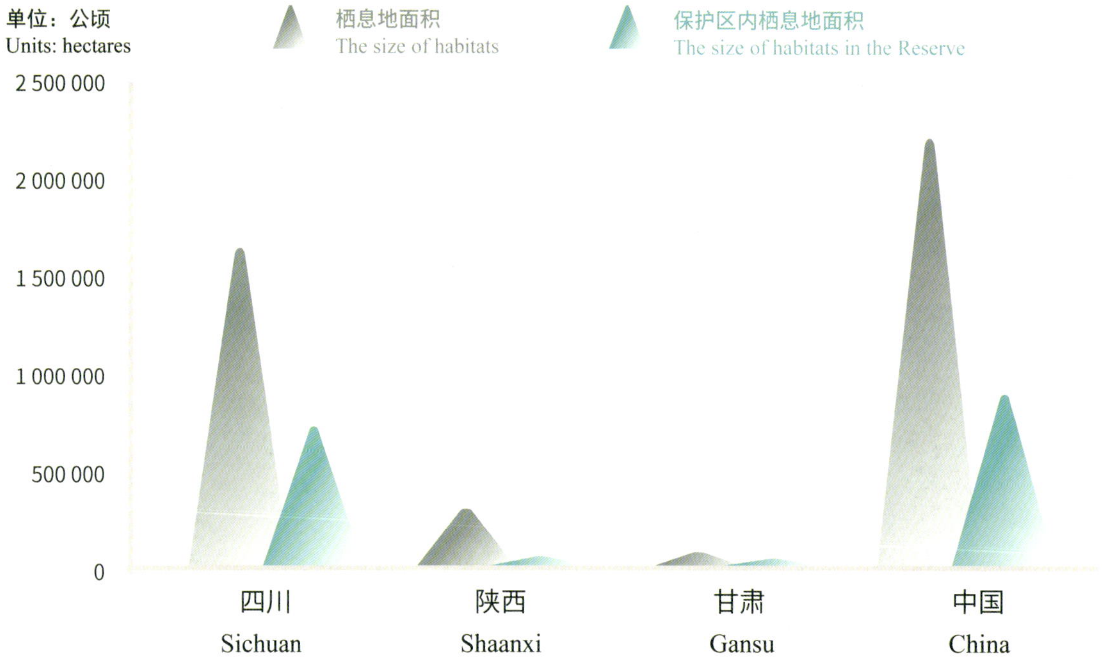

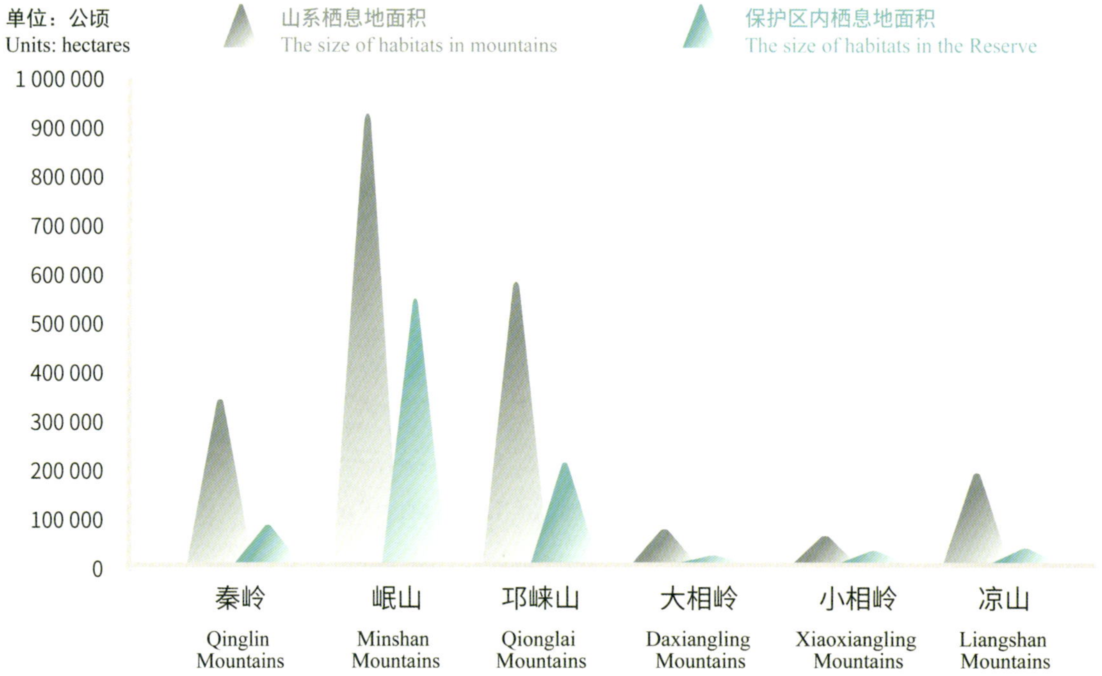

上／下 自然保护区内栖息地与大熊猫栖息地面积以及所在山系大熊猫栖息地面积比较图（2019 年）考拉看看 制图

资料来源：四川省地方志编纂委员会编《四川省志・大熊猫志》，方志出版社，2018 年

(Upper/Below) Comparison between the size of panda's habitat and that of the reserves as well as comparison between the size of habitat in the mountains and that in the reserves (2019). By Koalacan. Cited from *Sichuan Province Annals: The Giant Panda* compiled by Sichuan Local Records Editorial Committee, and published by Publishing House of Local Records, 2018

总部设在卧龙国家级自然保护区境内的中国大熊猫保护研究中心（2016 年） 周孟棋 摄

China Conservation and Research Center for the Giant Panda, Headquartered at Wolong National Nature Reserve (2016). Photo by Zhou Mengqi

栖息地面积

全国第四次大熊猫调查数据显示，大熊猫栖息地总面积约 2.58 万平方千米。与第三次调查数据相比，增幅为 11.8%。其中，四川省大熊猫栖息地面积占栖息地总面积的 78.7%，占比最大。

Habitat Area

According to China's Fourth National Census of the Giant Panda, its habitat totals about 25,800km², an increase of 11.8% compared to data from the third census, the largest being in Sichuan, accounting for 78.7% of the total.

栖息地竹资源

全国第四次大熊猫调查数据显示，中国大熊猫栖息地内共有 8 属 38 种竹种分布。竹种分布区域面积约 2.3 万平方千米，占栖息地总面积的 89.2%。大熊猫主食竹中分布面积较大的竹种有冷箭竹、秦岭箭竹、缺苞箭竹、巴山木竹等。

Bamboo Resources in the Habitat

The statistics from China's Fourth National Census of the Giant Panda indicate that there are 8 genera, 38 species of bamboo in total in the habitat, covering over 23,000km², 89.2% of the total area. The panda's staple edible bamboo species, which covers a larger area, include *Bashania fangiana, Fargesia qinlingensis, Fargesia denudata, Bashania fargesii* and so on.

2006 年 7 月 12 日，在立陶宛首都维尔纽斯召开的第 30 届世界遗产大会通过大熊猫栖息地遗产方案 蒲正祥 摄

On July 12, 2006, the 30th session of the UNESCO World Heritage Committee in Vilnius, capital of Lithuania, decided to put giant panda habitat in Sichuan on the World Heritage List as a natural heritage. Photo by Pu Zhengxiang

世界自然遗产

2004 年，中国正式向联合国教科文组织申报“四川大熊猫栖息地”为世界自然遗产项目。2006 年 7 月，世界遗产委员会通过集体表决，“四川大熊猫栖息地”被列入世界自然遗产名录。

World Natural Heritage

In 2004, China officially applied to the UN for listing Sichuan Giant Panda Sanctuaries in the World Heritage List and was consented in July 2006 by collective voting.

UNITED NATIONS EDUCATIONAL,
SCIENTIFIC AND
CULTURAL ORGANIZATION

CONVENTION CONCERNING
THE PROTECTION OF THE WORLD
CULTURAL AND NATURAL
HERITAGE

The World Heritage Committee
has inscribed

Sichuan Giant Panda Sanctuaries

on the World Heritage List

Inscription on this List confirms the exceptional and universal value of a cultural or natural site which requires protection for the benefit of all humanity

DATE OF INSCRIPTION
16 July 2006

DIRECTOR-GENERAL
OF UNESCO

2006 年 7 月，世界遗产委员会颁发的大熊猫栖息地遗产证书 孙前 供图

Giant Panda Habitat Heritage Certificate issued by World Heritage Committee in July 2006. Photo courtesy of Sun Qian

第四章 放归

如何才能让大熊猫这一濒临灭绝的物种在地球上壮大种群、繁衍生息？科研人员在人工繁殖饲养大熊猫的基础上，开展了圈养大熊猫野化培训与野外放归研究。

20 世纪 80 年代，中国便开始尝试放归野生大熊猫。大熊猫“贝贝”“三山”“桦桦”“遥远”[1] 等作为尝试放归对象被放归野外，但均以失败告终。

到了 21 世纪，圈养大熊猫的野化培训与放归时机逐渐成熟。2005 年，野生雌性大熊猫“盛林 1 号”闯入闹市被成功救护。在恢复健康后，科研人员给它带上定期自动脱落的 GPS 颈圈，将其放归至龙溪—虹口国家级自然保护区，到 2006 年推测判断它已初步融入当地野生大熊猫社群，健康地生活在放归区域。

国画《悄悄话》，50cm×50cm（2019 年）黄渝文 作品

Chinese painting: *Chitchatting*, 50cm×50cm (2019). By Huang Yuwen

“盛林 1 号”放归之后，科研人员进行了又一次尝试。这一次是雄性大熊猫“祥祥”的野化培训和放归试验，这也是人工繁殖大熊猫的首次试验。2006 年 4 月 28 日，在“祥祥”基本适应了二期野化培训圈内的野化环境后，专家们将其正式放归到卧龙国家级自然保护区西南部。

但不幸的是，“祥祥”因与放归区域野生大熊猫争夺领地和食物而发生遭遇战，激烈打斗造成严重内外伤，于 2007 年 2 月 19 日被发现死亡。

“盛林 1 号”和“祥祥”的两次放归尝试，为大熊猫野化放归研究提供了宝贵经验。

随后，四川省继续开展人工圈养大熊猫的野化放归和大熊猫孤立小种群复壮研究。2009 年 4 月 29 日，科研人员将从邛崃山山系救助的大熊猫“泸欣”在小相岭山系的四川省石棉县栗子坪自然保护区易地放归自然。2014 年 3 月和 5 月，用红外相机拍摄的照片中有 6 张照片和 2 段视频均显示放归大熊猫“泸欣”带崽活动。科研人员在放归区域开展了为期 1 年多的 DNA 样本收集

[1] 其中，“贝贝”“桦桦”在放归野外后死亡，“三山”放归失败后回到卧龙大熊猫饲养场，“遥远”放归失败后被送往雅安碧峰峡野生动物园饲养。

双面绣《梅花双熊》，40cm×40cm（2010 年） 郝淑萍 作品

Double-faced embroidery: *Two Giant Pandas Under A Plum Tree*, 40cm×40cm (2010). By Hao Shuping

蜀绣《小萌猫》，90cm×62cm（2017 年） 钟明 作品

Sichuan embroidery: *Adorable Baby Panda*, 90cm×62cm (2017). By Zhong Ming

和遗传分析工作，最终确定照片中的幼崽大熊猫为“泸欣”与当地野生大熊猫 LZP54（研究编号）的后代。这标志着，大熊猫小种群复壮进入一个新的阶段。

在过去的 10 年中，大熊猫野化放归工作稳步前进，建成中国大熊猫保护研究中心核桃坪野化培训基地、中国大熊猫保护研究中心天台山野化培训基地、成都大熊猫繁育研究基地都江堰野化培训基地 3 个野化培训基地，以及小相岭、大相岭大熊猫放归基地。从 2005 年到 2018 年，共野化培训了 14 只圈养大熊猫，其中 11 只放归自然，2 只死亡，其余全部存活；还将 2 只野外救护的大熊猫异地放归自然。

至此，人类在大熊猫野化放归研究上已取得长足进展。而这一研究还要继续下去，未来的路依旧漫长。

1

2

1. 水彩画《熊宝乐园》，30cm×21cm（2019 年） 四川省直属机关红星幼儿园 刘钰佳 作品 指导老师 张敏
2. 纸黏土画《快乐的一家》，30cm×22cm（2019 年） 四川省直属机关红星幼儿园 崔书瑜 作品 指导老师 杨倩
3. 水彩画《好好耍哦》，38cm×26cm（2019 年） 四川省直属机关红星幼儿园 唐煊智 作品 指导老师 杨倩
4. 水彩画《熊猫咪咪》，30cm×29cm（2019 年） 四川省直属机关红星幼儿园 边琪媛 作品 指导老师 王雪晶
5. 油水分离画《熊猫节看烟花》，26cm×38cm（2019 年） 四川省直属机关红星幼儿园 刘钰鑫 作品 指导老师 杨倩

Chapter 4 Reintroduction

To enable the giant panda to survive and expand its population, researchers have carried out studies on the reintroduction of captive pandas to the wild and on survival training as well.

As early as in the 1980s, China started the attempts to reintroduce captive giant pandas to the wild. However, all failed as shown by Bei Bei, San Shan, Hua Hua, and Yao Yuan[1].

The 21st century has proved to be the ripe time to give proper survival training to captive giant pandas and then reintroduce them to the wild. In 2005, Shenglin 1, a female giant panda from the wild, was captured when breaking into the downtown area of Dujiangyan City. Sick and wounded, it was taken good care of in captivity. After its recovery, researchers placed a GPS tracker collar (which will drop of its own in time) around its neck and released in Longxi-Hongkou National Nature Reserve. In 2006, it was estimated that it had integrated into the local wild panda population and lived a healthy life.

After this successful attempt, researchers took another try. This time they would reintroduce Xiang Xiang, a male giant panda. It was China's first artificially bred one to be released into the wild. On April 28, 2006, after Xiang Xiang had adapted to the wild environment in the second-stage training field, experts released it in the southwest of Wolong National Nature Reserve.

Unfortunately, it was found dead on February 19, 2007, from serious internal and external injuries caused by a fight for territory and food with some other wild panda.

[1]Bei Bei and Hua Hua died in the wild, San Shan came back to the feeding lot of Wolong, and Yao Yuan was recaptured and kept at Bifengxia Wildlife Park as it could not adapt to the wild.

1.Watercolor painting: *Cubs' Garden*, 30cm×21cm (2019). By Liu Yujia, from Hongxing kindergarten, under the instruction of Zhang Min

2.Paper clay painting: *Happy Family*, 30cm×22cm (2019). By Cui Shuyu, from Hongxing kindergarten, under the instruction of Yang Qian

3.Watercolor painting: *Having Fun*, 38cm×26cm (2019). By Tang Xuanzhi, from Hongxing kindergarten, under the instruction of Yang Qian

4.Watercolor painting: *Panda Mimi*, 30cm×29cm (2019). By Bian Qiyuan, from Hongxing kindergarten, under the instruction of Wang Xuejing

5.Oil-water separation painting: *Watching Fireworks During a Panda Festival*, 26cm×38cm (2019). By Liu Yuxin, from Hongxing kindergarten, under the instruction of Yang Qian

Failure in the two attempts gave precious lessons as to the giant panda reintroduction.

Subsequently, reintroduction efforts of the captive giant panda and expansion of the small isolated panda population were continued in Sichuan. On April 29, 2009, Lu Xin, a panda rescued from Qionglai Mountains, was released to the wild in Liziping National Nature Reserve in Shimian County, Sichuan. In March and May 2014, infrared cameras shot six photos and two videos showing Lu Xin nursing her baby. After more than a year of DNA sample collection and genetic analysis in the wild, researchers finally confirmed that the cubs in the photos were offspring of Lu Xin and a local wild giant panda LZP54 (research number). This means a new step of expanding the small isolated panda population by translocation release.

Over the past ten years, steady progress has been made in the reintroduction of giant pandas in to the wild; 3 survival training bases have been established: Hetaoping Survival Training Base and Tiantai Mountain Survival Training Base (both belonging to China Conservation and Research Center for the Giant Panda), and Dujiangyan Survival Training Base (belonging to Chengdu Research Base of Giant Panda Breeding); Giant panda reintroduction bases have also been set up in Xiaoxiangling Mountains and Daxiangling Mountains. From 2005 to 2018, 14 captive pandas received survival training, and 11 of them were reintroduced to nature—2 were found dead later. 2 giant pandas that had been rescued from the wild have also be reintroduced to localities other than their native habitat.

By now, great progress has been made in reintroducing the giant panda to the wild. But there is still a long way to go.

大熊猫幼崽开始尝试走路 引自四川省档案局、四川省档案学校编《档案见证·大熊猫》，四川人民出版社，2018 年

A panda cub is trying to walk. Cited from *Archives Witness: The Giant Panda* compiled by Sichuan Province Archives Bureau and Sichuan Archives College, and published by Sichuan People's Publishing House, 2018

2006 年 4 月 28 日，大熊猫“祥祥”在卧龙国家级自然保护区放归野外 四川省林业和草原局 供图

Xiang Xiang was being reintroduced into Wolong National Nature Reserve on April 28, 2006. Photo courtesy of Sichuan Province Forestry and Grassland Administration

2017 年 11 月 23 日上午，大熊猫“八喜”“映雪”在四川石棉县栗子坪国家级自然保护区内同时放归大自然进入山林生活 石棉县地方志工作办公室 供图

Ba Xi and Ying Xue were reintroduced into the Liziping National Nature Reserve on November 23, 2017. Photo courtesy of Shimian County Local Records Office

“祥祥”的野化培训

2003 年 7 月 8 日，大熊猫“祥祥”被放入一期小型野化培训圈。经过 15 个月的一期培训，“祥祥”基本适应了野化培训圈内的野化环境。2004 年 9 月 15 日，“祥祥”被转至二期野化培训圈。这一阶段的培训使“祥祥”在取食、警戒等方面更为适应自然，基本具备了在野外生存的能力。

Xiang Xiang’s Survival Training

On July 8, 2003, Xiang Xiang was put in the first stage small training field. After a 15-month survival training, it had been practically acclimated to the reintroduction environment. Then, on September 15, 2004, it was transferred to the second-stage training field, where it improved in foraging and vigilance, getting itself fully prepared to live in the wild.

2009 年 4 月 29 日，大熊猫“泸欣”在栗子坪自然保护区放归野外 石棉县地方志工作办公室 供图

Lu Xin was reintroduced into the Liziping National Nature Reserve on April 29, 2009. Photo courtesy of Shimian County Local Records Office

“泸欣”放归

“泸欣”是中国四川省首例易地放归的大熊猫，经工作人员长期跟踪观察后确认，“泸欣”已于 2012 年 8 月产下一崽。根据相关数据判断，“泸欣”所产幼崽为雄性，产崽地为“泸欣”放归地栗子坪自然保护区。这是首例产崽的野化放归大熊猫。

Lu Xin’s Release

Lu Xin’s reintroduction is the first case in Sichuan where a giant panda is released to an environment other than its native place. It gave birth to a baby in August 2012. Observation and collected data indicate that the baby is a male, born in Liziping National Nature Reserve where Lu Xin had been released. It is the first released panda that has given birth in the wild.

“淘淘”在三期野化培训场（2012 年） 衡毅 摄

Tao Tao in Phase III Field Training Ground (2012). Photo by Heng Yi

“淘淘”放归

“淘淘”是首只经母兽带崽野化培训后，放归自然的人工繁育大熊猫。它从小跟随妈妈“草草”，在自然环境下学习各种生存技能。2012 年 10 月 11 日，两岁的“淘淘”在栗子坪自然保护区放归自然，使得小相岭大熊猫小种群复壮试验更进一步。2017 年 12 月 28 日，经中国大熊猫保护研究中心的科研工作者确认，“淘淘”已基本适应了野外环境，健康状况良好。

Tao Tao's Release

Tao Tao was the first artificially bred giant panda that received survival training in the company of its mother Cao Cao. It learnt various survival skills from its mother in a natural environment. On October 11, 2012, it was reintroduced at the age of two to the wild in Liziping National Nature Reserve. On December 28, 2017, researchers in the China Conservation and Research Center for the Giant Panda confirmed that Tao Tao had practically acclimated itself to the wild environment.

第五章　保护力量

为保护包括大熊猫在内的野生动物，并给保护工作提供有力的组织保障，中国付出了巨大努力。

中华人民共和国成立以来，从中央到地方，建立了众多保护机构，逐渐形成了完善的保护体系。

1950 年 5 月，中国发布法规，开始对大熊猫和森林等资源展开保护与管理。20 世纪 50 年代，大熊猫的保护工作由政府多个部门共同管理。为更好地保护野生动物与限制狩猎，国务院于 1958 年和 1962 年两次下发通知，要求各地迅速将这一工作统一交由林业部门管理，并加强有关管理机构[1]的建设。

2018 年 10 月 29 日下午，大熊猫国家公园管理局在成都揭牌。2018 年 11 月，大熊猫国家公园四川、陕西、甘肃三省管理局陆续挂牌。2019 年 1 月 15 日，大熊猫国家公园四川省管理局成都市、绵阳市、雅安市、广元市、阿坝藏族羌族自治州、德阳市、眉山市 7 个管理分局在相关市（州）同时举行挂牌仪式。

大熊猫的保护力度日渐加大。大熊猫国家公园的建立，对于推进自然资源科学保护和合理利用，促进人与自然和谐共生，推进美丽中国建设，具有极其重要的意义。同时，随着大熊猫保护工作的推进，相关保护机构和体系建设逐渐完善。大熊猫保护工作取得巨大成就，与保护机构和体系的建立密不可分。

1

2

国画《大熊猫》系列
直径 42cm（2018 年）　成都市龙江路小学
1. 李玮索菲　作品
2. 刘冷云曦　作品
3. 李姝贤　作品
4. 郭若水　作品
指导老师　余文兰

[1] 引自 1962 年国务院发布的《关于积极保护和合理利用野生动物资源的指示》。

Chapter 5 Conservation Institutions

3

4

Chinese painting: *The Giant Panda Series*, 42cm in diameter (2018).

1. By Li Wei Suo Fei

2. By Liu Leng Yun Xi

3. By Li Shuxian

4. By Guo Ruoshui

from Chengdu Longjiang Road Primary School, under the instruction of Yu Wenlan

China has made great efforts to protect the giant panda and other animals, which is a powerful organizational support for protection.

Since the founding of the People's Republic of China in 1949, a good number of institutions have been established by the central and local authorities, thus forming a well-organized protection system.

In May 1950, China initiated its protection and management of the giant panda and forest resources. In the 1950s, its protection was jointly managed by different governmental departments. To better protect wildlife and to restrict hunting, the State Council demanded in 1958 and 1962 that giant panda protection and management be handed over to government organizations in charge of forestry and construction of panda protection institutions be strengthened.[1]

On October 29, 2018, Giant Panda National Park Administration was inaugurated in Chengdu. In November 2018, Giant Panda National Park Administration was established respectively in Sichuan, Shaanxi and Gansu. On January 15, 2019, seven branch administrations in Chengdu, Mianyang, Ya'an, Guangyuan, Aba, Deyang and Meishan, were unveiled.

The establishment of the national park plays a significant role in the protection and utilization of natural resources, in harmonious coexistence of man and nature, and in the landscape rehabilitation. As the giant panda conservation progresses, protection institutions and systems have been established and optimized. Till now, tremendous achievements have been made in giant panda conservation, and credit for the achievements should go to these institutions and systems.

[1]From *Directives on Conservation and Rational Utilization of Wildlife Resources* issued by the State Council of China in 1962.

1989 年施行的《中华人民共和国野生动物保护法》

Law of The People's Republic of China on the Protection of Wildlife which came into force in 1989.

《中华人民共和国野生动物保护法》

这是一部保护和拯救珍稀、濒危野生动物，保护、发展和合理利用野生动物资源，维护生态平衡的法律。该部法律于 1988 年 11 月 8 日经第七届全国人民代表大会常务委员会第四次会议通过。2018 年 10 月，第十三届全国人民代表大会常务委员会第六次会议予以修订。

Law of the People's Republic of China on the Protection of Wildlife

This is a law concerning rare, endangered wildlife protection, wildlife resources development and utilization, and ecological balance maintenance. It was passed at the Fourth Session of the Standing Committee of the Seventh National People's Congress of the People's Republic of China on November 8, 1988. In October 2018, the Sixth Session of the Standing Committee of the Thirteenth National People's Congress of the People's Republic of China decided to make an amendment to it.

左 2007 年 8 月，中国野生动物保护协会与澳大利亚阿德莱德动物园达成协议，共同实施为期 10 年的大熊猫繁殖研究计划。图为“网网”“福妮”赴澳欢送仪式 引自四川省档案局、四川省档案学校编《档案见证·大熊猫》，四川人民出版社，2018 年

右 1984 年 4 月，美国总统夫人南希·里根在美国发起“零钱给熊猫”募捐活动，将募集的 13077.44 美元转交中国野生动物保护协会。图为位于四川省宝兴县蜂桶寨国家级自然保护区的南希·里根为抢救大熊猫捐款纪念碑 引自四川省档案局、四川省档案学校编《档案见证·大熊猫》，四川人民出版社，2018 年

(Left) In August 2007, the China Wildlife Conservation Association reached an agreement with the Adelaide Zoo in Australia to jointly carry out a 10-year panda breeding research program. The picture shows the farewell ceremony. Cited from *Archives Witness: The Giant Panda* compiled by Sichuan Province Archives Bureau and Sichuan Archives College, and published by Sichuan People's Publishing House, 2018

(Right) In April 1984, Nancy Davis Reagan, first lady of the United States, launched a fundraising campaign named "Pennies for Pandas" in the United States. The total donation of $13,077.44 was given to the China Wildlife Conservation Association. This is the memorial monument in honor of Nancy Reagan's contribution to giant panda protection in Fengtongzhai National Nature Reserve, Baoxing County, Sichuan Province.Cited from *Archives Witness: The Giant Panda* compiled by Sichuan Province Archives Bureau and Sichuan Archives College, and published by Sichuan People's Publishing House, 2018

中国野生动物保护协会

1983 年 12 月 23 日，中国野生动物保护协会在北京成立。其宗旨是推动中国野生动物保护事业与社会经济的可持续发展，促进人与自然和谐共生。中国野生动物保护协会是中国最大、具有广泛代表性的生态保护组织，拥有会员 41 万多人。

China Wildlife Conservation Association

Established in Beijing on December 23, 1983, China Wildlife Conservation Association aims to promote wildlife conservation and sustainable social and economic development in China, and to promote harmony and coexistence between man and nature. The Association, with over 410,000 members, acts as the largest and most representative ecological conservation organization in China.

1990 年，陕西省汉中地区野生动物保护站的科研人员和当地群众救助大熊猫“梨梨” 新华社 供图

Local people and scientists at the Hanzhong Wildlife Conservation Station in Shaanxi Province rescued giant panda Li Li in 1990. Photo courtesy of Xinhua News Agency

2015—2016 年，四川省凉山彝族自治州冕宁县林业局野生动物保护站在全县开展大熊猫保护巡护宣传 余祖静 摄

From 2015 to 2016, the Wildlife Protection Station of Mianning County Forestry Administration in Liangshan Yi Autonomous Prefecture, Sichuan Province carried out a county-wide publicity campaign for giant panda protection. Photo by Yu Zujing

2015 年 9 月 23 日，村民巡护队队员在甘肃省白水江国家级自然保护区巡护 新华社 供图
Members of the village patrol team were patrolling Baishuijiang National Nature Reserve in Gansu Province on September 23, 2015.
Photo courtesy of Xinhua News Agency

川陕甘保护行动

大熊猫的栖息地分布在中国四川、陕西和甘肃三省，从 20 世纪 50 年代开始，川陕甘三省针对大熊猫的保护工作日益加强，从研究到保护，逐渐组建了专业机构。在三地的大熊猫保护工作中，市县及基层也建立了很多保护大熊猫等野生动物的机构。

Conservation in Sichuan, Shaanxi and Gansu

As giant pandas inhabit Sichuan, Shaanxi and Gansu, the governments in the three provinces have been strengthening their protection since the 1950s. Research and conservation institutions have been set up at municipal, prefectural and local levels.

第六篇
文化

蜀绣《亲子》，100cm×50cm（2018 年） 孟德芝 作品

Sichuan embroidery: *Mother and children*, 100cm×50cm (2018). By Meng Dezhi

Part 6
Culture

它是一种符号，是人类对美的定义

在地球上，还不曾有过像大熊猫一样受人追捧的动物。

它以独特的气质和迷人的外形，俘获了全人类的芳心。

它创造了一种风靡全球的美学，

150 年来人类对于大熊猫的热爱与日俱增。

The giant panda is a symbol that stands for human definition for beauty

Never have there been creatures on this planet as popular with human beings as the giant panda.

Its distinctive and adorable appearance touches every human heart

and creates a kind of aesthetics prevailing over the world,

for human beings have been increasingly fond of it ever since its discovery over 150 years ago.

凝视（2014 年） 张志和 摄
Gaze (2014). Photo by Zhang Zhihe

2010 年 12 月 18 日，成都市民欢送“开开”“心心”飞赴澳门 成都大熊猫繁育研究基地 供图
On December18, 2010, Chengdu residents bid farewell to Kai Kai and Xin Xin when they were flying to Macau. Photo courtesy of Chengdu Research Base of Giant Panda Breeding

第一章 探亲

2006 年，中国农历新年，全球观众通过中国中央电视台春节联欢晚会的平台，为两只大熊猫投票选出了乳名。它们作为祖国母亲的礼物，赠送给宝岛台湾。雄性大熊猫叫“团团”，雌性大熊猫叫“圆圆”。

在准备了近 3 年之后，2008 年 12 月 23 日，“团团”“圆圆”离开故土四川赶赴台湾。而几个月之前，“团团”和“圆圆”亲历了“5·12”汶川特大地震，幸运的是，它们在巨大滚石砸到圈舍之前成功逃生获救。

这两只同甘共苦的大熊猫感情笃深。2013 年 7 月，“团团”和“圆圆”产下首胎千金，取名“圆仔”，这只幼崽一出生便成为大明星。

除台湾之外，香港和澳门也曾获赠大熊猫。1999 年，在香港回归祖国两周年之际，“安安”和“佳佳”从内地到香港安家。2007 年和 2009 年，香港和澳门回归十周年之际，两地都曾获得来自中央人民政府赠送的大熊猫。

这是探亲之礼，更是中华民族大家庭团结的象征。

版画《熊猫》 吴冠中 作品 引自孙前著《大熊猫文化笔记》，五洲传播出版社，2009 年
Printmaking: *Giant Pandas*, By Wu Guanzhong. Cited from Sun Qian, *Notes of the Giant Panda Culture*, China Intercontinental Press, 2009

Chapter 1 Gifts from Home

At the CCTV Spring Festival Gala in 2006, audience all over the world voted pet names for a pair of giant pandas for Taiwan as gifts from the Motherland. The male was later named Tuan Tuan and the female Yuan Yuan.

On December 23, 2008, they left their homeland Sichuan for Taiwan. Seven months before, on May 12, a violent earthquake had struck their habitat in Wenchuan. Fortunately, they had been rescued before huge falling stones smashed their enclosure.

They mated and gave birth to a female cub in July 2013, which was named Yuan Zai and became a superstar in Taiwan right after.

Just as Taiwan has received giant pandas as gifts, so have Hong Kong and Macao. In 1999, two giant pandas, An An and Jia Jia, were sent to Hong Kong as gifts for the 2nd anniversary of its return to China. On their 10th anniversaries of return to the Motherland (2007 for Hong Kong and 2009 for Macao), both Hong Kong and Macao received giant pandas as gifts from the Central People's Government for these special occasions.

They are not only gifts from home but also the symbol of the unity of the Chinese nation.

阳光正好（2013 年） 杨老丹 摄
Warm sunshine (2013). Photo by Yang Laodan

“团团”和“圆圆”在吃竹子（2008 年） 谭楷 供图
Tuan Tuan and Yuan Yuan are nibbling at bamboos (2008). Photo courtesy of Tan Kai

“团团”和“圆圆”的新家

台北市立动物园是“团团”和“圆圆”的新家，动物园为它们设立了网站，还开通了博客，向民众展示它们的日常生活。动物园还为“团团”和“圆圆”准备了丰盛的台湾竹子。至 2009 年，两只大熊猫所食用的竹子全部产自台湾。此外，台湾还发行了大熊猫“团团”“圆圆”邮票。

New Home for Tuan Tuan and Yuan Yuan

Taipei Zoo, a new home for Tuan Tuan and Yuan Yuan, created a website for them and started a blog to display their daily activities to the Taiwan compatriots, who had Taiwan bamboo planted specially for them. By 2009, all the bamboo the two giant pandas ate was produced locally. Taiwan also issued stamps featuring Tuan Tuan and Yuan Yuan.

2016年8月8日，大熊猫“佳佳”过38岁生日，成为“迄今为止最长寿的圈养大熊猫” 引自四川省档案局、四川省档案学校编《档案见证·大熊猫》，四川人民出版社，2018年

On August 8, 2016,Giant panda Jia Jia celebrated her 38th birthday and became “the oldest captive giant panda ever”. Cited from *Archives Witness: The Giant Panda* compiled by Sichuan Province Archives Bureau and Sichuan Archives College, and published by Sichuan People’s Publishing House, 2018

吉尼斯世界纪录

2015年7月28日，大熊猫“佳佳”在香港即将迎来37周岁生日。当天，“佳佳”刷新了当时的两项吉尼斯世界纪录，成为“迄今为止最长寿的圈养大熊猫”及“最长寿的在世圈养大熊猫”。

Guinness World Records

On July 28, 2015, giant panda Jia Jia, having been living in Hong Kong and soon to turn 37, broke two Guinness World Records as “the oldest captive giant panda ever” and “the oldest living captive giant panda.”

左 2016 年 5 月 11 日，由香港特区政府援建的中华大熊猫苑举行开园仪式 引自四川省档案局、四川省档案学校编《档案见证·大熊猫》，四川人民出版社，2018 年

右 香港特区政府援建的中国大熊猫保护研究中心卧龙神树坪基地中的大熊猫（2017 年） 朱强 摄

(Left) The China Giant Panda Park, built in Hong Kong, was holding its opening ceremony on May 11, 2016. Cited from *Archives Witness: The Giant Panda* compiled by Sichuan Province Archives Bureau and Sichuan Archives College, and published by Sichuan People's Publishing House, 2018

(Right) Giant pandas in Wolong Shenshuping Base of China Conservation and Research Center for the Giant Panda, built by Hong Kong SAR Government (2017). Photo by Zhu Qiang

香港援建卧龙

2008 年“5·12”汶川特大地震给中国卧龙国家级自然保护区带来巨大损失，与大熊猫故乡结下深厚情谊的香港特别行政区政府决定援建卧龙保护区及相关设施，共计 23 个灾后重建项目，援建资金 14.22 亿元人民币。

Wolong Nature Reserve Aided by Hong Kong SAR Government

Owing to the great damage of the 2008 Wenchuan Earthquake to Wolong National Nature Reserve, Hong Kong Special Administrative Region decided to help rebuild it and its affiliated facilities, which totals an involvement of 23 projects, and investment of ¥1.422 billion.

我美不？（2016 年） 周孟棋 摄
Am I Beautiful? (2016) Photo by Zhou Mengqi

2001 年，四川出入境检验检疫局工作人员对出境大熊猫进行检疫 成都海关 供图

Staff of Sichuan Inspection and Quarantine Bureau were quarantining the outbound giant panda in 2001. Photo courtesy of Sichuan Inspection and Quarantine Bureau

第二章 出国

在被科学发现的早期，西方博物学家、探险家以捕猎等方式将大熊猫带出中国。

这一时期出现过两次大熊猫热潮。第一次是大熊猫刚刚被科学发现后，即清同治八年（1869年），在科学上确认大熊猫是一个新物种，在西方社会引起轰动。

第二次是在民国 25 年（1936 年），美国女服装设计师露丝•哈根纳斯第一次将活体大熊猫“苏琳”带出中国。一经展出，“苏琳”立即成为最受美国人民喜爱的动物明星，展出第一天观众就达 4 万人。

民国 25 年至民国 35 年（1936—1946 年），共有 16 只大熊猫通过不同途径被运往西方国家。至 1953 年 9 月，美国的最后一只大熊猫“梅梅”去世后，整个西方国家便再也没有活体大熊猫存在。

1939 年，“潘多拉”飞向大洋彼岸 谭楷 供图
Giant panda Pandora flew to the other side of the Pacific Ocean in 1939. Photo courtesy of Tan Kai

Chapter 2 Going Abroad

After the French missionary Armand David first discovered the giant panda in 1869, western naturalists and explorers came to China to hunt and smuggle them out of China.

This period witnessed two giant panda crazes. The first occurred shortly after its discovery in 1869, when identified as a new species in zoology, causing a sensation in the western world.

The year 1936 marked the second craze when Ruth Harkness, an American female fashion designer, took giant panda Su Lin alive out of China for the first time. As soon as it was exhibited, Su Lin became the most popular "animal star" in America. On the very first day of its exhibition, the audience amounted to 40,000.

From 1936 to 1946, 16 giant pandas were taken to western countries in different ways. By September 1953, when Mei Mei, the last giant panda, died in the United States, there had been no longer a living giant panda in the entire western world.

国画《天伦之乐 国之瑰宝》，138cm×68cm（2018 年） 夏国栋 作品

Chinese painting: *Family Happiness, National Treasure,*138cm×68cm (2018). By Xia Guodong

女皇赠“白熊”

据日本《皇家年鉴》记载，唐代垂拱元年（685 年）10 月，女皇武则天曾将一对称为“白熊”的活体动物和 70 张毛皮作为大唐珍贵的国礼赠送给日本天武天皇。此“白熊”被研究者认为是大熊猫。

“White Bears” as Presents from the Empress

According to Japan's *Royal Almanac*, in October 685, Empress Wu Zetian in the Tang Dynasty, presented Japanese Emperor Tenmu with 70 pelts and a pair of living “white bears”, which are believed to be giant pandas according to researchers.

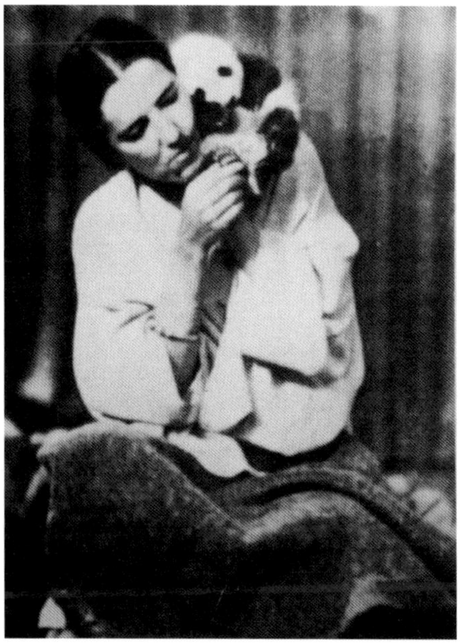

20 世纪 30 年代，露丝和“苏琳” 谭楷 供图

Ruth and Su Lin in the 1930s. Photo courtesy of Tan Kai

大熊猫“苏琳”

大熊猫被科学发现后，很多外国人纷纷前往中国试图捕获大熊猫。民国 25 年（1936 年）9 月 26 日，美国女服装设计师露丝哈根纳斯为继承亡夫动物学家威廉哈根纳斯遗志，几经辗转，终于来到中国四川省威州（今汶川县）寻找大熊猫。11 月 9 日，当露丝和她的同行者一起上山查找捕捉大熊猫的猎套时，发现了一只大熊猫幼崽。这只幼崽便是大熊猫“苏琳”。

Giant Panda Su Lin

After the discovery of the giant panda, hunters from other countries came to track them. On September 26, 1936, in order to fulfill the unfinished work of her late zoologist husband, William Harkness, Ruth Harkness, an American female fashion designer, came to search for giant pandas in Weizhou (today's Wenchuan County), Sichuan. On November 9, when she and her companions went up to the mountains checking the trap set for giant pandas, they found a cub and named it Su Lin.

宋美龄和纽约动物园代表在一起 薛康 供图

Song May-ling and Representatives from New York Zoo. Photo courtesy of Xue Kang

宋美龄赠美大熊猫

为感谢“美国援华联合会”对中国抗日战争的援助，民国 30 年（1941 年），宋美龄以个人名义赠送给美国“潘弟”和“潘达”两只大熊猫。由于当时第二次世界大战还未结束，它们需要乘飞机先从香港到菲律宾，再乘船到夏威夷，最后抵达美国旧金山，合计绕路约 56326 千米。

Giant Pandas to the US by Song May-ling

In the thirtieth year of the Republic of China (1941), to express her gratitude to the United China Relief for its support in China's War of Resistance Against Japan, Song May-ling presented a pair of giant pandas, Pandee and Pandah. Since World War II was still going on, the pandas flew from Hong Kong to the Philippines first, then travelled by ship to Hawaii and finally arrived in San Francisco, a distance totaling 56,326km.

英国摄影师 Bert Hardy 的儿子与大熊猫“明”在一起 引自谭楷著《熊猫的故事》，五洲传播出版社，2015 年

The son of British photographer Bert Hardy and the giant panda Ming. Cited from Tan Kai, *Story of the Giant Panda*, China Intercontinental Press, 2015

无畏的“明”

民国 25 年至民国 27 年（1936—1938 年），英国人丹吉尔 · 史密斯高价收购 12 只大熊猫，并成功将 6 只运回英国。其中一只叫“明”的大熊猫最逗人喜爱。在第二次世界大战期间，当法西斯的铁蹄践踏西欧，德国战机轰炸伦敦之时，“明”依然闲庭信步，玩耍自如，成为英勇无畏的象征，鼓舞着英国人民奋勇抵抗。

Fearless Ming

From 1936 to 1938, Tangier Smith from Britain, purchased 12 giant pandas with a large sum of money and managed to transport 6 of them to UK. One of them, Ming, was the cutest and most adorable one. During World War II, when German jet fighters bombarded London, Ming kept enjoying itself, in spite of the circumstances, and was regarded as a symbol of courage and fearlessness, inspiring the British people to fight on.

甘冽清甜，真爽！（2017 年） 张志和 摄
Sweet and cool! (2017) Photo by Zhang Zhihe

第三章　外交

从20世纪50年代起，大熊猫作为象征和平与友好之礼，开始走出国门。中国政府先后将24只大熊猫作为国礼赠送给苏联、美国、日本、英国、法国等9个国家。当时，到中国访问的国家元首和政府首脑都把获赠大熊猫当作特殊的“政治待遇”，这让大熊猫成为团结世界人民的友谊大使。

1972年2月，美国总统尼克松访华。在尼克松举行的答谢宴上，国务院总理周恩来指着熊猫牌香烟盒上的熊猫图案，向尼克松夫人帕特·尼克松表示将向美国赠送两只大熊猫。尼克松夫人听后十分惊喜。《华盛顿邮报》发表评论说：“周恩来通过可爱的大熊猫，一下子把美国人的心征服了。”

大熊猫“玲玲”和“兴兴”作为友谊使者于1972年4月抵达美国，入驻华盛顿国家动物园。这对大熊猫开放参观后，每年大约有300万人前往观看。

日本也非常喜欢大熊猫，曾通过各种渠道，多次向中国表达希望获赠大熊猫的愿望。1972年，中国赠送给日本两只大熊猫“兰兰”和“康康”。这两只大熊猫得到了日方极高的礼遇，日本人民更是对大熊猫偏爱有加，每天有1.46万名观众专程到东京上野动物园，只为一睹大熊猫芳容。短短几年之内，参观者就有5000多万人次。

20世纪80年代后，为保护大熊猫，中国政府决定停止对外赠送。为了满足各国人民观赏大熊猫的热切心愿，发展同世界人民的友谊，中国开始同有关国家进行大熊猫国际合作研究。

水墨画《九寨藏珍》
97cm×180cm（2007年）
邱笑秋 作品

Chapter 3 Diplomacy

and colour painting: *Jewels in zhaigou*, 97cm×180cm (2007). By Qiu Xiaoqiu

Since the 1950s, the giant panda, a token of peace and friendship, has begun to travel abroad. The Chinese Government presented 24 giant pandas as national gifts to 9 countries including the Soviet Union, the United States, Japan, Britain and France. At that time, a gift of the giant panda was a special "political treatment" for heads of states and governments visiting China. So, the giant panda has assumed the role as China's goodwill ambassadors.

At the returning banquet for U.S. President Nixon and his wife on their visit to China in February 1972, Premier Zhou Enlai, pointing at the panda pattern on a cigarette box, told the First Lady that two giant pandas were to be presented to the United States. And the other was more than happy at that. *The Washington Post* commented that "Zhou Enlai has won the American heart with a cute giant panda."

As goodwill ambassadors, giant pandas Ling Ling and Hsing Hsing arrived in the United States in April 1972 and were settled in the National Zoological Park. About 3 million people visited them every year after the two pandas were on display.

The Japanese people also longed for the giant panda. The Government there repeatedly expressed this wish to its Chinese counterpart through various channels. In 1972, giant pandas Lan Lan and Kang Kang were sent there and became so popular that they had 14,600 visitors at Ueno Zoo, Tokyo every day. In few years, visitors exceeded 50 million.

In the 1980s, in order to better protect the giant panda, the Chinese Government decided to stop sending them as gifts. But to satisfy people's keen desire and promote friendship, China began to conduct research of the giant panda in cooperation with other countries.

20 世纪 50—80 年代赠送外国大熊猫一览表[1]
Table of Giant Pandas Presented to Foreign Countries from the 1950s to the 1980s[1]

赠送时间 Year	获赠国家 Recipient	大熊猫 Name	来源 From
1955	苏联 Soviet Union	姬姬、平平 Chi Chi, Ping Ping	四川宝兴 Baoxing, Sichuan
1957	苏联 Soviet Union	安安 An An	四川宝兴 Baoxing, Sichuan
1964	朝鲜 Korea	1 号、2 号 Number 1, Number 2	四川宝兴 Baoxing, Sichuan
1970	朝鲜 Korea	凌凌、三星 Lin Lin, San Xing	四川宝兴 Baoxing, Sichuan
1972	美国 USA	玲玲、兴兴 Ling Ling, Hsing Hsing	四川宝兴 Baoxing, Sichuan
1972	日本 Japan	兰兰、康康 Lan Lan, Kang Kang	四川宝兴 Baoxing, Sichuan
1973	法国 France	黎黎、燕燕 Li Li, Yan Yan	四川宝兴、四川平武 Baoxing and Pingwu, Sichuan
1974	英国 Britain	晶晶、佳佳 Jing Jing, Chia Chia	四川平武、四川宝兴 Pingwu and Baoxing, Sichuan
1975	墨西哥 Mexico	贝贝、迎迎 Pe Pe, Ying Ying	四川越西、四川宝兴 Yuexi and Baoxing, Sichuan
1978	西班牙 Spain	绍绍、强强 Shao Shao, Qiang Qiang	北京动物园 Beijing Zoo
1979	朝鲜 Korea	丹丹 Dan Dan	四川宝兴 Baoxing, Sichuan
1980	联邦德国 Federal Germany	天天、宝宝 Tian Tian, Bao Bao	四川天全、四川宝兴 Tianquan and Baoxing, Sichuan
1980	日本 Japan	欢欢 Huan Huan	四川宝兴 Baoxing, Sichuan
1982	日本 Japan	飞飞 Fei Fei	四川 Sichuan

黎黎、燕燕 Li Li, Yan Yan

姬姬 Chi Chi

平平 Ping Ping

丹丹 Dan Dan

迎迎、贝贝 Ying Ying, Pe Pe

[1] 考拉看看 制图 资料来源：四川省地方志编纂委员会编《四川省志·大熊猫志》，方志出版社，2018 年

[1] By Koalacan. Cited from *Sichuan Province Annals: Giant Panda* compiled by Sichuan Local Records Editorial Committee, and published by Publishing House of Local Records, 2018

晶晶、佳佳
ing Jing, Chia Chia

兰兰、康康 Lan Lan, Kang Kang

天天 Tian Tian

宝宝 Bao Bao

安安 An An

飞飞 Fei Fei

玲玲、兴兴 Ling Ling, Hsing Hsing

欢欢 Huan Huan

三星 San Xing

凌凌 Lin Lin

1 号 Number 1

2 号 Number 2

强强、绍绍 Qiang Qiang, Shao Shao

20 世纪 50—80 年代，中国赠送外国的 24 只大熊猫 马德里动物园、宝兴县林业局、四川省档案馆、新华社、宝兴县地方志工作办公室、史密森学会、考拉看看，高华康等 供图

giant pandas presented to foreign countries from the 1950s to the 1980s. Photo courtesy of Zoo Aquarium de Madrid, Baoxing County Forestry Bureau, Sichuan Province Archives, Xinhua News Agency, Baoxing County Local Records, Smithsonian Institution, Koalacan, and Gao Huakang

大熊猫到过的国家（1936—2005 年）

爱尔兰	法国	瑞典	奥地利	韩国
泰国	澳大利亚	荷兰	西班牙	比利时
加拿大	新加坡	朝鲜	美国	新西兰
德国	墨西哥	英国	俄罗斯	日本

考拉看看 制图 资料来源：赵学敏编《大熊猫——人类共有的自然遗产》，中国林业出版社，2006 年

Countries which giant pandas visited（1936-2005）				
Ireland	France	Sweden	Austria	South Korea
Thailand	Australia	Netherlands	Spain	Belgium
Canada	Singapore	Korea	USA	New Zealand
Germany	Mexico	Britain	Russia	Japan

By Koalacan. Cited from Zhao Xuemin, *The Giant Panda: Natural Heritage of Humanity*, China Forestry Publishing House, 2006

走向世界，受人追捧 周孟棋 摄

Favorite of the whole world. Photo by Zhou Mengqi

1. 成都 IFS 大厦的巨型大熊猫雕塑 (2016 年) 袁博 摄

2. 成都红星路步行街夜景下的大熊猫雕塑 (2017 年) 袁博 摄

3. 大熊猫展舍前每天人潮涌动，从各地来的游客争先一睹大熊猫的风采 谭楷 供图

4. 1984 年，洛杉矶动物园， 参观

大熊猫的观众排起了长队 谭楷 供图

1.Giant Panda Sculpture in Chengdu IFS Building (2016). Photo by Yuan Bo

2.A night view of the giant panda sculpture in the pedestrian street on Red Star Road, Chengdu (2017). Photo by Yuan Bo

3.Tourists from all over the world flocking to giant panda houses to see the cute pandas every day. Photo courtesy of Tan Kai

4.Visitors lining up to see giant pandas at the Los Angeles Zoo in 1984. Photo courtesy of Tan Kai

让世界沸腾的大熊猫

1973 年 12 月 8 日，大熊猫"黎黎""燕燕"落户巴黎文森动物园。当时整个巴黎为之沸腾，争相参观"中国派来的动物大使"，就连大熊猫雕塑和照片都被抢购一空，可谓盛况空前。而墨西哥为了庆祝大熊猫幼崽"多威"的出生，在墨西哥儿童节这天，政府组织了声势浩大的彩车游行活动，场面颇为壮观。

Giant Pandas That Excited the World

On December 8, 1973, giant pandas Li Li and Yan Yan got settled in the Zoo de Vincennes. The whole city of Paris was excited. People flooded there for a look at the "Animal ambassadors from China." It was an unprecedented event: even the giant panda sculptures and photos were sold out. In order to celebrate the birth of giant panda cub Tohui in Mexico, the Mexican Government organized a large-scale, spectacular floral parade on the Children's Day.

山姆鹰的客人

1984 年 7 月，在邓小平同志过问下，一对四川籍大熊猫"永永"和"迎迎"做客洛杉矶奥运会。大熊猫在奥运会展出期间，观众场场爆满，前后共有 200 万人前去参观。美国人民称大熊猫是"尊贵使者""山姆鹰[1]的客人"。

Guests of Sam the Olympic Eagle

In July 1984, Deng Xiaoping requested that giant pandas Yong Yong and Ying Ying, born in Sichuan, be sent to the Los Angeles Olympic Games as visiting guests, which enjoyed a sell-out exhibition with two million visitors in total. Americans called them "Honorable envoys" and "Guests of Sam the Olympic Eagle[1]".

[1] 洛杉矶奥运会吉祥物为象征美国的"山姆鹰"。

[1]Mascot for Los Angeles Olympics and symbol of the United States.

水中照样行 引自魏辅文著《野生大熊猫科学探秘》，科学出版社，2018 年
Wading the stream. Cited from Wei Fuwen, *Scientific Exploration of Wild Giant Pandas*, Science Press, 2018

第四章 符号

1961 年 9 月 11 日，世界野生动植物基金会（WWF，简称“基金会”）在瑞士成立。在此之前，大熊猫“姬姬”在伦敦动物园展出，出现了万人空巷的盛况。对此，基金会创始人英国生物学家、曾任联合国教科文组织总干事的朱安利·赫胥黎以及鸟类学家、曾任英国自然保护组织总干事的马克斯·尼科尔森十分惊叹。他们意识到这只来自东方的迷人动物对人们有着巨大的号召力。它不仅珍贵，其黑白相间的独特外形更是一个醒目的标志。

基金会最早的标志草稿由苏格兰自然学家杰拉尔德·华生起草，后来，基金会第一届主席彼得·斯科特在此基础上以大熊猫“姬姬”为主体设计了基金会标志。此后的 40 多年中，基金会的标志几经调整，不断改进，但那只憨态可掬的大熊猫始终是标志的主体，并成为基金会的象征。随着基金会全球会员队伍日益壮大，这个“大熊猫标志”开始在越来越多的国家成为自然保护运动的标志。

大熊猫的形象后来又出现在亚运会和奥运会上。不仅如此，大熊猫在全世界成了一种常见的、充满美好寓意的符号。中国野生动物保护协会的会徽，是一个抽象的大熊猫图案；成都的市徽图案以大熊猫啃食竹叶为主体；2006 年、2007 年，中俄两国政府互办了“国家年”，双方都把代表中国的大熊猫和代表俄罗斯的棕熊两种动物手牵手的图案作为两国“国家年”的标志；此外，把大熊猫的形象注册为商标的企业更是数不胜数……

在全球掀起热潮的大熊猫，不再只是一种珍稀动物，它已经成为象征珍贵、友谊、和平的文化符号，并深深铭刻在全世界人民心中。

世界自然基金会（WWF）LOGO 演变 引自孙前著《大熊猫文化笔记》，五洲传播出版社，2009 年

LOGO Evolution of WWF. Cited from Sun Qian, *Notes of the Giant Panda Culture*, China Intercontinental Press, 2009

Chapter 4 Symbol

生活在伦敦动物园的大熊猫"姬姬"，是世界自然基金会（WWF）会徽的原型 引自四川省林业厅编《图说大熊猫》，天地出版社，2015 年

Chi Chi, a giant panda living in London Zoo, is the prototype of the WWF emblem. Cited from *The Giant Panda Pictures* compiled by Forestry Department of Sichuan Province and published by Tiandi Press, 2015

薄胎提花竹编瓶《国宝》，直径 16cm，高 36cm（1988 年） 刘嘉峰 作品

Egg-shell bamboo woven bottle: *National Treasure*, 16cm in diameter, 36cm in height (1988). By Liu Jiafeng

On September 11, 1961, World Wide Fund for Nature (WWF) was founded in Switzerland. Prior to that, giant panda Chi Chi had been exhibited at the London Zoo, which made such a sensation that thousands of people turned out for him, which so amazed Julian Huxley and Max Nicholson, the former being a founding member of WWF, British biologist and former Director-General of UNESCO, and the latter an ornithologist and former Director-General of the British Association of Nature Conservationists, that they realized that the fascinating oriental creature was appealing because of its unique black-and-white appearance, which could serve as an eye-catching logo.

The first logo for WWF was drafted by Gerald Watterson, a Scottish naturalist. Later, Peter Scott, WWF's first president, designed its logo, based on Chi Chi. In the following four decades, the logo had been constantly improved, but the charming panda was always there as a symbol of the Fund. With the growing membership of WWF, the "giant panda logo" has become a symbol of the nature conservation movement in more and more countries.

The image of the giant panda later appeared in both Asian Games and Olympic Games as a token of goodwill. The emblem for China Wildlife Conservation Association is an abstract pattern of a giant panda; the logo of Chengdu features a giant panda eating bamboo leaves; Chinese and Russian governments hosted "China Year in Russia" and "Russia Year in China" respectively in 2006 and 2007, both taking the pattern of a Chinese giant panda hand in hand with a Russian brown bear as the logo on the special occasions. Besides, countless companies have registered the image of the giant panda as their trademarks.

The global giant panda crave made this species no longer a rare creature but a cultural symbol in terms of rarity, friendship and peace.

大熊猫文化形象嬗变

大熊猫在古时被认为是一种猛兽，甚至有古籍记载，它在驯养后被应用于战争。这一时期，大熊猫被赋予了“战神”和“保护神”的文化寓意。但是现在，大熊猫圆滚滚的形态及其食竹的习性，让它成为象征和平、友好的文化形象。

Evolution of Giant Panda as a Cultural Image

Records of the giant panda can be found in ancient Chinese books. Regarded as a ferocious beast in ancient times, it was used in warfare after domesticated. At that time, the giant panda was given the cultural meanings of “God of War” and “Patron Saint”. But now, the round-shaped and bamboo-eating creature has become a symbol of peace and friendship.

雪地太极（2017 年） 周孟棋 供图

Playing Taiji in the snow (2017).
Photo courtesy of Zhou Mengqi

1. 2005 年 11 月 11 日，第 29 届北京奥运会吉祥物发布仪式在北京举行 谭楷 供图

2. 2008 年第 29 届北京奥运会吉祥物之一——“福娃晶晶”，代表奥林匹克五环中黑色的一环 引自赵学敏编《大熊猫——人类共有的自然遗产》，中国林业出版社，2006 年

3. “盼盼”标志 引自赵学敏编《大熊猫——人类共有的自然遗产》，中国林业出版社，2006 年

4. 2015 年 9 月 12 日，由国家体育总局、中国残疾人联合会主办，四川省人民政府承办的中华人民共和国第 9 届残疾人运动会暨第 6 届特殊奥林匹克运动会开幕式在四川省成都市举行 四川省残疾人联合会 供图

5. 1990 年，“巴斯”被亚洲运动会组织委员会选中，并以“盼盼”为名成为亚洲运动会吉祥物 引自四川省档案局、四川省档案学校编《档案见证·大熊猫》，四川人民出版社，2018 年

6. 1991 年，大熊猫“盼盼”在中国中央电视台春节联欢晚会上 引自四川省档案局、四川省档案学校编《档案见证·大熊猫》，四川人民出版社，2018 年

1		4	
2	3	5	6

1.On November 11, 2005, the mascot launching ceremony of the 29th Olympic Games was held in Beijing. Photo courtesy of Tan Kai

2.Giant panda Jing Jing, one of the mascots of the 29th Summer Olympic Games held in Beijing in 2008, represents the black one of the Olympic rings. Cited from Zhao Xuemin, *The Giant Panda: Natural Heritage of Humanity*, China Forestry Publishing House, 2006

3.The mascot for Pan Pan. Cited from Zhao Xuemin, *The Giant Panda: Natural Heritage of Humanity*, China Forestry Publishing House, 2006

4.The 9th Disabled Games of the People's Republic of China and the 6th Special Olympics, sponsored by the General Administration of Sport of China and the China Disabled Persons' Federation and undertaken by the People's Government of Sichuan Province, were launched on September 12, 2015 in Chengdu, Sichuan Province. Photo courtesy of Sichuan Disabled Persons' Federation

5.Ba Si has been chosen by the Asian Games Organizing Committee and named Pan Pan as the mascot of the Asian Games in 1990. Cited from *Archives Witness: The Giant Panda* compiled by Sichuan Province Archives Bureau and Sichuan Archives College, and published by Sichuan People's Publishing House, 2018

6.In 1991, the mascot or Pan Pan at the Spring Festival Gala of CCTV. Cited from *Archives Witness: The Giant Panda* compiled by Sichuan Province Archives Bureau and Sichuan Archives College, and published by Sichuan People's Publishing House, 2018

大熊猫从中国走向世界，“熊猫文化世界共享”——2018年8月，首届中国大熊猫国际文化周在北京举行，四川航空公司空客A350“大熊猫客机”首航 高华康 摄

Giant pandas going out of China and "Sharing Panda Culture with the World"—the 1st China Giant Panda International Culture Week was held in Beijing in August 2018. Sichuan airlines Airbus A350 named "Giant Panda Passenger Plane" about to make its maiden flight. Photo by Gao Huakang

2015 年，中国大熊猫主题文化艺术展览“黑与白的魔法时空——熊猫艺术”在澳门举行 成都大熊猫繁育研究基地 供图

Chinese Giant Panda Theme Culture and Art Exhibition: “*Black and White Magic Time and Space—Panda Art*” was held in Macau in 2015. Photo courtesy of Chengdu Research Base of Giant Panda Breeding

上：四川航空（2017 年）；下：成都地铁
号线的熊猫元素（2016 年） 周孟棋 供图

与大熊猫相关的纪录片（2019 年） 考拉看看 制图

Documentaries about the giant panda (2019). By Koalacan

熊猫频道（2019 年） 考拉看看 制图

Ipanda (2019). By Koalacan

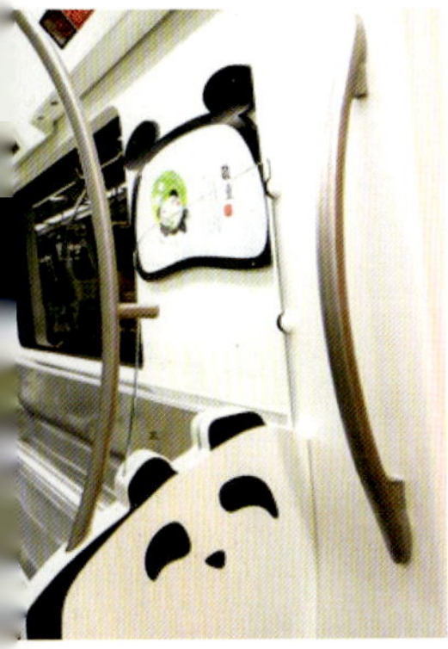

Upper) Sichuan Airlines 2017); (Below) Panda Train on hengdu Metro Line 3 (2016). hoto courtesy of Zhou Mengqi

中国人民银行于 1983 年 12 月 5 日发行银质熊猫纪念币，以纪念 1983 年被世界野生动植物基金会定为世界熊猫年。纪念币的图案正面为北京天坛祈年殿，背面为母子熊猫与青竹，每枚面值人民币 10 元 新华社 供图

The People's Bank of China issued silver panda commemorative coins on December 5, 1983 in honor of the World Panda Year designated by the WWF in 1983.On the face of the coin is the Hall of Prayer for Good Harvest in the Temple of Heaven in Beijing, while on its back are a mother panda and her child with bamboo behind. The face value of each silver coin is RMB 10. Photo courtesy of Xinhua News Agency

中国人民银行从 1982 年 9 月 15 日起，陆续在国外和中国香港等地发行中国熊猫金币。中国熊猫金币分 1 盎司、1/2 盎司、1/4 盎司和 1/10 盎司 4 个品种，成色均为 99.9%。金币正面图案为北京著名古建筑天坛祈年殿（右上），背面图案均为中国珍稀动物熊猫与青竹 新华社 供图

Since September 15, 1982, the People's Bank of China has been issuing China Panda gold coins in some foreign countries and Hong Kong, China. There are four types in terms of their face values: 1 ounce, 1/2 ounce, 1/4 ounce and 1/10 ounce, with gold purity of 99.9% in each. On the face side of each gold coin is the Hall of Prayer for Good Harvest in the Temple of Heaven in Beijing (upper left), while on the back the giant panda—Chinese rare animal—and bamboo. Photo courtesy of Xinhua News Agency

成都市市徽（1997 年后，该市徽不再使用）

Logo of Chengdu (No longer in use after 1997)

成都国际熊猫节节徽

Emblem of Chengdu International Panda Festival

雅安市市徽

Logo of Ya'an

第 6 届全国大学生运动会吉祥物“欣欣”

Xin Xin,mascot of 6th Universiade of People's Republic of China

第 2 届全国体育大会吉祥物“珂珂”

Ke Ke,mascot of 2nd National Sports Festival of China

大熊猫图标

以大熊猫为设计灵感的会徽或图标随处可见，并且受到全世界人民欢迎。除世界自然基金会的标识外，中国野生动物保护协会会徽、四川旅游标志、雅安市市徽等多种图标，均将大熊猫作为标志。

Giant Panda as a Logo

The giant panda is so popular around the world that it is frequently used as the logo of organizations such as WWF, China Wildlife Conservation Association, and Sichuan Tourism. And it is taken as the emblem by Ya'an City, too.

左／右　中俄“国家年”标志

(Left/Right) Logo of “China Year in Russia” and “Russia Year in China”

2017 年成都国际马拉松赛标志

Logo of 2017 Chengdu International Marathon

中国野生动物保护协会（CWCA）会徽

Emblem of China Wildlife Conservation Association (CWCA)

2018 年首届中国国际进口博览会标志

Emblem of 1st China International Import Expo,2018

四川旅游标志

Logo of Sichuan Tourism

中国成都 2019 年第 18 届世界警察和消防员运动会标志（左）和吉祥物（中、右）
Logo (left) and mascot (middle, right) of 2019 Chengdu World Police & Fire Games

2019 年，成都熊猫亚洲美食节吉祥物 Panda 胖墩
Panda Pang Dun, Mascot of Chengdu Panda Asian Food Festival 2019

2019 年，四川省文化和旅游发展大会标志
Logo of Culture and Tourism Development Conference of Sichuan Province, 2019

2019 年 6 月，“阿璞”被评选为首个大熊猫国际形象。图为“阿璞”国际系列形象的基础形象 新华社 供图

A Pu, a cartoon image of a young male panda, was chosen through public appraisal as the first global representative of China’s giant pandas in June 2019. It is the basic image of A Pu global series. Photo courtesy of Xinhua News Agency

“阿璞”衍生形象“音乐有我”

A Pu’s derivative image: Musician

“阿璞”衍生形象“公益有我”

A Pu’s derivative image: Public Welfare Activist

“阿璞”衍生形象“环保有我”

A Pu’s derivative image: Environmentalist

1993 年成都国际熊猫节 周孟棋 供图

Chengdu International Panda Festival in 1993. Photo courtesy of Zhou Mengqi

国际熊猫节

中国四川成都国际熊猫节先后于 1993 年和 1997 年举行。国际熊猫节的标志图案是悠闲吃竹叶的大熊猫坐在一只大手上，寓意大熊猫是人类的“掌上明珠”，所有人都应该认识到它的珍贵，积极参与保护大熊猫、保护环境的活动。

International Panda Festival

The International Panda Festival was held in Chengdu, China, respectively in 1993 and 1997. Its logo was a giant panda seated on a large palm relishing bamboo leaves, implying that the giant panda is the "pearl in my palm" (equivalent to the English proverb "apple of my eye"). This logo is aimed to help people realize the preciousness of giant pandas and play an active part in protecting them and nature.

左 / 右上 / 右下 成都市熊猫路小学及其文创作品 成都市熊猫路小学 供图

(Left/Upper right/Lower right) Chengdu Panda Road Primary School and its students' works. Photo courtesy of Panda Road Primary School

成都市熊猫路小学

因与成都大熊猫繁育研究基地毗邻而得名。学校与成都大熊猫繁育研究基地、成都动物园建立友好合作关系，每年开展的“熊猫艺术节”“熊猫体育节”“熊猫读书节”“熊猫科技月”等活动使学生身心得到全面发展，“熊猫文化、自然教育”逐渐成为学校文化的灵魂。2019 年 4 月 23 日，国内第一座以大熊猫为主题的图书馆在成都市熊猫路小学内开馆。

Chengdu Panda Road Primary School

The School is thus named for its location adjacent to Chengdu Research Base. It has established friendly cooperative relations with Chengdu Panda Base and Chengdu Zoo. It holds panda-related activities every year, such as “Panda Art Festival” “Panda Sports Festival” “Panda Reading Festival” and “Panda Sci-Tech Month,” which has not only enabled its pupils to develop physically and academically, but also provided chances for them to receive “nature education through panda culture.” On April 23, 2019, China's first panda-themed library opened in the School.

熊猫邮局

熊猫邮局是中国邮政于 2013 年 10 月宣布成立的全新文创子品牌，这是全球首个以熊猫为主题的邮局。它是中国邮政打造的传播书信文化和熊猫文化的空间。

熊猫邮局少城路旗舰店（2016 年） 熊猫邮局 供图

Flagship panda post on Shaocheng Road, Chengdu, Sichuan Province (2016). Photo courtesy of Panda Post

Panda Post Office

Founded in October 2013, the Panda Post Office is a newly innovated sub-brand of China Post as the world's first of its kind and a platform for spreading the correspondence culture and panda culture.

航空
中国人民邮政
1963
8－5
北京
熊猫
挂号
0940

168
Giant Panda, Chicago
中国人民邮政
PHOTO BY ACME NEWSPICTURES, INC.

150. GEBURTSTAG CARL HAGENBECK
150 JAHRE BERLINER ZOO
100
CARL HAGENBECK 1844–1913
DEUTSCHE BUNDESPOST
200
150 JAHRE BERLINER ZOO
DEUTSCHE BUNDESPOST

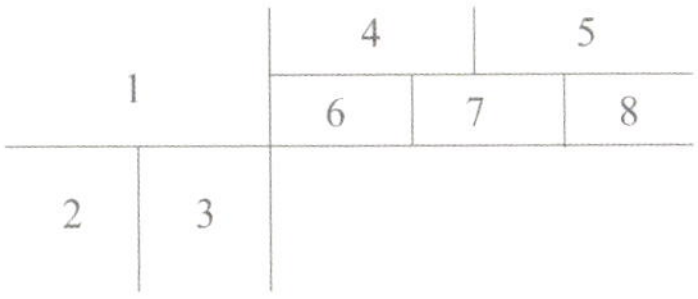

1. 邮票实寄封——1963年，中国发行世界第一套以熊猫为主图的邮票，图为由北京寄青岛的航空挂号封 薛康 供图

2. 邮票极限明信片——1936年，大熊猫“苏琳”在美国布鲁克菲尔德动物园展出时，由芝加哥斯托尔公司发行的世界第一批熊猫明信片（贴世界第一套以熊猫为主图的邮票，盖第一个熊猫邮局的首日邮戳）薛康 供图

3. 邮票小全张——1994年，柏林动物园150周年邮票小全张 薛康 供图

4. 邮票小型张——中非共和国为香港2015邮票展览会发行的以大熊猫为背景的小熊猫邮票小型张（异形）薛康 供图

5. 邮票小全张——2011年，几内亚比绍共和国发行的中国大熊猫邮票小全张（异形）薛康 供图

6. 邮票小型张首日封——2013年，朝鲜熊猫邮票小型张首日封 薛康 供图

7. 邮票首日封——1982年，日本野生动物邮票首日封 薛康 供图

8. 长隆野生动物世界为大熊猫三胞胎制作《萌帅酷成长记》纪念册 薛康 供图

1.A used envelope with stamps—first set of panda stamps issued in China in 1963. This is a registered airmail from Beijing to Qingdao. Photo courtesy of Xue Kang

2.Maximum cards—the world's first panda postcards released by the J. O. Stoll Company in Chicago in 1936 when giant panda Su Lin was exhibited at the Brookfield Zoo in the United States. They bear the world's first set of panda stamps and the first-day postmarks of the first giant panda post office. Photo courtesy of Xue Kang

3.A mini sheet—for the 150th anniversary of Berlin Zoo, 1994. Photo courtesy of Xue Kang

4.A souvenir sheet: Red Panda stamp, issued by Central African Republic for Hong Kong 2015 Stamp Exhibition (irregular shape). Photo courtesy of Xue Kang

5.A mini sheet—China Giant Panda Minisheet (irregular shape) issued by the Republic of Guinea in 2011. Photo courtesy of Xue Kang

6.A souvenir sheet first-day cover—2013 North Korean Giant Panda First Day Cover. Photo courtesy of Xue Kang

7.A first-day Cover—1982 Japanese Wildlife Stamp first-day Cover. Photo courtesy of Xue Kang

8.A memoir by Guangzhou Chimelong Safari Park named *Growth Diary* for the giant panda triplets. Photo courtesy of Xue Kang

沐浴阳光的大熊猫（2009 年） 张志和 摄
Basking in the sun (2009). Photo by Zhang Zhihe

第五章 海归

按照《濒危动植物物种国际贸易公约》和中国政府与他国政府签订的有关协议，参与国际合作的大熊猫的所有权全部属于中国，合作期满全部归国；倘若死亡，尸体也归属中方。合作期间的大熊猫产下幼崽，所有权归中国。

2018 年夏天，旅居奥地利的雄性大熊猫“龙徽”在离乡 15 年后回到了中国。遗憾的是，它是在去世后被制作成标本送回祖国的。

“龙徽”出生于 2000 年，在 3 岁的时候与雌性大熊猫“阳阳”一起前往奥地利美泉宫动物园。在 16 年的时间里，这一对熊猫夫妇自然交配相继生育了“福龙”“福虎”“福豹”“福凤”“福伴”5 胎，创造了欧洲圈养大熊猫自然交配生产的纪录。

2016 年 12 月，“龙徽”因癌症救治无效死亡，享年 16 岁（相当于人类 60 岁）。经中奥双方协商，奥方将“龙徽”的尸体制作成标本，运返中国，用于大熊猫科研和公众教育。

2010 年 11 月 19 日，全球首个“海归大熊猫乐园”在中国保护大熊猫研究中心雅安碧峰峡基地成立。开园当日，有 6 只“海归”大熊猫住在了这里，分别是“苏琳”“珍珍”“华美”“美生”“泰山”以及“龙徽”的大儿子“福龙”。

2018 年 12 月 4 日，随着“龙徽”和“阳阳”的龙凤胎宝宝“福凤”“福伴”回国，已有 15 只大熊猫回到中国大熊猫保护研究中心生活。至此，“龙徽”的 5 个宝宝全部回到祖国。

所有回国的大熊猫都需要经历 1 个月的隔离检疫，确保健康后，才会与公众见面。此外，它们还要度过几个月到半年时间不等的适应期，以适应家乡生活。这些旅居国外的大熊猫二代回国后，将担负起繁殖后代的重要任务。

1

2

3

Chapter 5 Overseas Returnees

中国大熊猫保护研究中心雅安碧峰峡基地“海归大熊猫乐园” 引自四川省档案局、四川省档案学校编《档案见证·大熊猫》，四川人民出版社，2018 年

“ Oversea-born Panda Paradise” at Ya’an Bifengxia Base, China Conservation and Research Center for Giant Panda. Cited from *Archives Witness: The Giant Panda* compiled by Sichuan Province Archives Bureau and Sichuan Archives College, and published by Sichuan People’s Publishing House, 2018

1. 油画《I love my big Bamboo》 80cm×80cm（2016 年）
2. 油画《It’s so yummy》 80cm×80cm（2016 年）
3. 油画《I am coming》 80cm×80cm（2016 年）

安东尼奥·威利 作品

1. Oil painting: *I love my big Bamboo*, 80cm×80cm (2016) .
2. Oil painting: *It’s so yummy*, 80cm×80cm (2016).
3. Oil painting: *I am coming*, 80cm×80cm (2016).

By Antonio Wehrli

According to the *Convention on International Trade in Endangered Species of Wild Fauna and Flora* (CITES) and the relevant agreements signed by the Chinese Government with its cooperative agencies abroad, the giant panda to be engaged in any cooperative project is exclusively owned by China and must be returned as the agreement expires. In case of death, its carcass belongs to China. And all cubs born during the cooperation belong to China, too.

In the summer of 2018, male giant panda Long Hui, a sojourner in Austria for 15 years, returned to China, but unfortunately, in the form of a specimen.

Long Hui was born in 2000 and, at age three, went to the Tiergarten Schönbrunn in Austria with female giant panda Yang Yang. In the span of 16 years, the couple, through natural mating, gave birth to five cubs (Fu Long, Fu Hu, Fu Bao, Fu Feng and Fu Ban), setting a record of natural mating and reproduction of captive giant pandas in Europe.

In December 2016, it died of cancer at age 16 (equivalent to the human age of 60) after a medical treatment failure. Through negotiation, its body was made into a specimen and shipped back to China for scientific research and public education.

On November 19, 2010, the world’s first “Overseas-born Panda Paradise” was established in Ya’an Bifengxia Base of China Panda Center. On the opening day, six “returned pandas” were settled there, including Su Lin, Zhen Zhen, Hua Mei, Mei Sheng, Tai Shan and Fu Long (Long Hui’s eldest son).

On December 4, 2018, with the return of Fu Feng and Fu Ban, the pigeon pair by Long Hui and Yang Yang, 15 giant pandas lived in China Panda Center. So far, Long Hui’s 5 children have all been back in China.

Before meeting the public, all the returnees require a one-month quarantine to ensure their good health. In addition, they need time for adaptation to the environment in their home country, which may take months or even half a year. Back home, these second-generation giant pandas will assume the important task of reproduction.

大熊猫“海浜”（左）和“阳浜”（右）引自四川省档案局、四川省档案学校编《档案见证·大熊猫》，四川人民出版社，2018 年

Giant pandas Hai Bang (left) and Yang Bang (right). Cited from *Archives Witness: The Giant Panda* compiled by Sichuan Province Archives Bureau and Sichuan Archives College, and published by Sichuan People's Publishing House, 2018

海外最大的熊猫家族

1994 年，成都大熊猫繁育研究基地与日本和歌山县白浜野生动物园开始了大熊猫国际科研合作。经 20 多年的繁衍生息，日本和歌山县白浜野生动物园“浜”系家族成为中国大熊猫旅居海外最大的熊猫家族。截至 2017 年 6 月，已有 11 只大熊猫从日本回到中国。

Largest Panda Family Overseas

In 1994, Chengdu Panda Base and Wakayama Adventure World in Shirahama, Japan, began their cooperation on giant panda research. In the span of over 20 years' reproduction, the "Bang" family in the Wakayama Adventure World has become the largest among those sojourning overseas. As of June 2017, 11 giant pandas had left Japan for China.

2000 年 1 月 25 日，“华美”（左）在圣地亚哥动物园与游人见面；2004 年，5 岁的“华美”（右）回到中国四川卧龙 引自四川省档案局、四川省档案学校编《档案见证·大熊猫》，四川人民出版社，2018 年

Hua Mei was meeting visitors at the San Diego zoo on January 25, 2000 (left); 5-year-old Hua Mei returned to Wolong, Sichuan, China in 2004 (right). Cited from *Archives Witness: The Giant Panda* compiled by Sichuan Province Archives Bureau and Sichuan Archives College, and published by Sichuan People's Publishing House, 2018

第一只归国大熊猫

1999 年 8 月 21 日，雌性大熊猫“华美”在美国圣地亚哥动物园出生。2004 年 2 月 13 日，“华美”回到卧龙国家级自然保护区，成为全球第一只出生在国外且回到中国的大熊猫。截至 2012 年，“华美”共生育了 8 只大熊猫，是一位多产的母亲。

First Returnee

On August 21, 1999, female giant panda Hua Mei was born in the San Diego Zoo in USA. On February 13, 2004, it returned to Wolong National Nature Reserve, being the first overseas-born returnee. By 2012, this prolific mother had given birth to 8 cubs altogether.

龙凤胎回乡

2018 年 12 月 4 日，旅奥大熊猫龙凤胎“福凤”“福伴”回到中国，入住中国大熊猫保护研究中心卧龙神树坪基地隔离检疫区，成为首对“海归”大熊猫双胞胎。2016 年 8 月 7 日，它们在奥地利美泉宫动物园出生，“福凤”为雌性，“福伴”为雄性。

Pigeon Pair's Return Home

On December 4, 2018, Fu Feng (female) and Fu Ban (male), the pigeon pair sojourning in Austria, returned to China and stayed in the separated quarantine area at Shenshuping Panda Base, becoming the first "returned" twins. They were born in the Tiergarten Schönbrunn in Austria on August 7, 2016.

海归大熊猫“福豹”的幸福生活（2017年） 刘茜 摄

Happy life of giant panda returnee Fu Bao (2017). Photo by Liu Qian

2010 年 3 月 15 日，“海归”明星大熊猫“美兰”在成都大熊猫繁育研究基地迎来回国后与公众的首次见面 新华社 供图

Mei Lan, a well-known oversea-born giant panda, makes her debut at Chengdu Research Base of Giant Panda Breeding on March 15, 2010. Photo courtesy of Xinhua News Agency

旅美大熊猫重回故乡

2018 年 11 月 1 日，雄性大熊猫“高高”在旅美 15 年后回到中国，入住中国大熊猫保护研究中心都江堰基地。这是中美大熊猫科研合作协议期满后，重新回到故乡的首只大熊猫。回国时，“高高”已有 28 岁（相当于人类年龄约 98 岁）。回国后，基地为其定制了饲养护理方案，并安排专职人员进行饲养管理，使其能够安享晚年。

2019 年 5 月 16 日，旅美大熊猫“白云”“小礼物”搭乘航班回到了家乡四川，被送往中国大熊猫保护研究中心都江堰基地进行为期 30 天的隔离检疫。至此，“高高”“白云”“小礼物”一家又在四川故乡团聚。1996 年 9 月，母亲“白云”被中国保护大熊猫研究中心送往美国圣地亚哥动物园开展国际合作，已阔别家乡 23 年。2012 年，“白云”在圣地亚哥动物园产下雄性大熊猫“小礼物”。

Back Home after Sojourn in USA

On November 1, 2018, male giant panda Gao Gao returned to China after 15 years' stay in the United States and settled in Dujiangyan Giant Panda Base of China Conservation and Research Center. This is the first that has returned as the Sino-US giant panda research cooperation agreement came to an end. Upon its return, it was 28 years old (equivalent to 98 of human age). Since then, the Base has made a special feeding and nursing program supervised by a caretaker to ensure its comfort and peace in old age.

On May 16, 2019, Xiao Liwu and his mother Bai Yun, came back by air to China from the United States, and were sent to the Dujiangyan Base of China Conservation and Research Center for the Giant Panda for a 30-day quarantine. As this point, the family of Gao Gao, Bai Yun and Xiao Liwu were reunited in their homeland—Sichuan. It had been 23 years since Bai Yun was sent in September 1996 to the San Diego Zoo on an international cooperation project by the China Conservation and Research Center for the Giant Panda. In 2012, Bai Yun gave birth to a male cub at the San Diego zoo, named Xiao Liwu (Little Gift in English).

第六章　传播

大熊猫是大自然的馈赠，也是全人类共同的财富。它的生存与保护不仅是全世界生物学界关注的焦点，也牵动着全球人民的心。在大熊猫保护事业上，不断努力探索的不止有科研人员，普通大众也在贡献着自己的力量。多年来，全球有不计其数的个人和团队，捐赠资金用于抢救、保护大熊猫等活动，同时还积极对熊猫文化进行传播与推广。

1999 年伊始，为纪念逝去的父亲，美国熊猫国际的执行董事苏珊·布兰登开始认养以其父亲名字命名的大熊猫“Harvey（哈维）”。2009 年 5 月 12 日，影视艺人成龙认养大熊猫“龙龙”“凤凤”。

类似的案例还有很多。不少跨国公司、社会组织、企业集团、知名人士等纷纷参与其中，通过爱心认养的方式表达对大熊猫的珍爱之情。

除了爱心认养外，不少名人还通过担任“熊猫大使”来宣传保护大熊猫。2009 年，成龙受聘为成都熊猫大使，成为全球首位熊猫大使。2010 年和 2015 年，纪录片《马文的熊猫探险记》制片人奈杰尔·马文、俄罗斯国宝级歌唱家瓦列莉娅，也分别被授予“熊猫大使”称号。2019 年，美国驻成都总领事馆总领事林杰伟成为熊猫公益推广志愿者。

爱心认养大熊猫、出任“熊猫大使”，虽然只是大熊猫保护工作中的一小部分，但是产生了深远的意义。它传播了熊猫文化，让更多社会人士了解大熊猫的生长环境、关注大熊猫的健康成长，进而号召越来越多的组织和个人加入到保护濒危动物的行列中来，共同保护人类赖以生存的美丽家园。

2006 年 9 月 29 日，澳大利亚籍教师安珍和认养的大熊猫“安珍”在一起　新华社　供图

Australian teacher Jaye and her adopted giant panda An Zhen, on September 29, 2006. Photo courtesy of Xinhua News Agency

Chapter 6 Publicity

上 20 世纪 80 年代，成都动物园饲养员正在给大熊猫“蓉蓉”喂食 肖龙联 摄

下 20 世纪 80 年代，北京动物园熊猫馆三位年轻的饲养员正在和大熊猫嬉戏 王德 摄

(Upper) Rong Rong was being fed at Chengdu Zoo in the 1980s. Photo by Xiao Longlian

(Below) In the 1980s, three young keepers in the Panda House of Beijing Zoo were frolicking with pandas. Photo by Wang De

The giant panda is a gift from Mother Nature and the shared treasure of mankind. Its survival and protection are not only the focus of the global biological community, but also the concern of the whole world. In the course of its protection, both researchers and commoners have been making their contributions. Over the years, countless individuals and organizations all over the world have donated for its rescue and conservation and have engaged in the publicity of panda culture.

At the beginning of 1999, Suzanne Braden, executive director of Pandas International, adopted giant panda Harvey, named after her late father. On May 12, 2009, famous film star Jackie Chan adopted two giant pandas—Long Long and Feng Feng.

There is no lack of similar cases. Many international corporations, organizations, conglomerates, and celebrities have been adopting giant pandas.

In addition, many celebrities are involved in making the giant panda public by serving as “Pambassadors.” In 2009, Jackie Chan was appointed Pambassador for Chengdu, as the world’s first pambassador. Nigel Marven, producer of the documentary *Panda Adventure with Nigel Marven*, and Valeria Pavlova, well-reputed Russian singer, were respectively honoured as “Pambassador” in 2010 and 2015. Jim Mullinax, U.S. Consulate General in Chengdu, has become a volunteer of giant panda promotion in 2019.

Although adopting a giant panda or being a “Pambassador” is a small part of the giant panda conservation, these activities have far-reaching significance. They have helped spread panda culture, help more people to understand the habitat environment of giant pandas and pay attention to pandas’ health, thereby appealing to more and more organizations or individuals to be involved in protecting endangered animals and our beautiful homeland.

卧龙大熊猫俱乐部组织认养人一年一度的"拥抱宝宝卧龙行"活动 引自四川省档案局、四川省档案学校编《档案见证·大熊猫》，四川人民出版社，2018 年

Wolong Giant Panda Club organizes the annual "Hug My Baby at Wolong" event for adopters; Cited from *Archives Witness: The Giant Panda* compiled by Sichuan Province Archives Bureau and Sichuan Archives College, and published by Sichuan People's Publishing House, 2018

卧龙大熊猫俱乐部

卧龙大熊猫俱乐部成立于 1992 年，多年来一直在组织大熊猫捐赠和认养活动。俱乐部利用多种渠道参与国内外交流与合作，先后与多个公益组织和新闻媒体建立了良好的合作关系，如美国熊猫国际、日本大熊猫保护协会、爱能博格基金会、日本大熊猫研究所等。卧龙大熊猫俱乐部在资助大熊猫科研项目、大熊猫栖息地保护、大熊猫饲养管理以及社区共建等方面做出了很多努力，发挥了重要作用。

Wolong Panda Club

Wolong Panda Club, set up in 1992, has been organizing activities of donation for and adoption of the giant panda by cooperating with various non-profit organizations and news agencies in and outside China, such as Pandas International, Giant Panda Protection Institute of Japan, Annenberg Foundation, and Giant Panda Research Institute of Japan. The Club has played a significant role in giant panda research projects, habitat protection, feeding and nursing programs.

美国熊猫国际公益慈善组织执行董事苏珊与她认养的大熊猫“哈维”的合影 引自四川省档案局、四川省档案学校编《档案见证·大熊猫》，四川人民出版社，2018 年

Suzanne Braden, executive director of Pandas International Charities, USA, and her adopted giant panda Harvey. Cited from *Archives Witness: The Giant Panda* compiled by Sichuan Province Archives Bureau and Sichuan Archives College, and published by Sichuan People's Publishing House, 2018

美国熊猫国际

美国熊猫国际是一个公益慈善组织，为中国保护大熊猫研究中心贡献了很多力量，如为大熊猫保护、科研、繁育、医疗保健、栖息地保护等事业募集资金。此外，还积极为来自世界各地的 60 多名认养人提供帮助，促进大熊猫认养活动。

Pandas International

As a charity organization, it has been supporting China Conservation and Research Center for the Giant Panda, through raising funds for its conservation, research, breeding, health care, and habitat protection. Also, it has helped more than 60 adopters from all over the world.

2014 年 7 月 10 日，抵达法国埃斯佩莱特市的罗维孝 高华康 供图

Luo Weixiao arrived at Espelette, France on July 10, 2014. Photo courtesy of Gao Huakang

熊猫骑士罗维孝

2014 年 3 月 18 日，64 岁的四川雅安市民罗维孝从雅安市宝兴县邓池沟天主教堂出发，经哈萨克斯坦、俄罗斯、白俄罗斯、波兰、德国等国，历时 106 天，骑行约 1.5 万千米，于 2014 年 7 月 10 日到达大熊猫科学发现者阿尔芒 · 戴维的故乡法国埃斯佩莱特市。罗维孝一路宣传大熊猫文化，被称为“CHINA 大熊猫文化骑士”。

Luo Weixiao the Panda Cyclist

Luo Weixiao, a 64-year-old citizen of Ya’an, starting from Dengchigou Catholic Church on March 18, 2014, Baoxing County, Ya’an, arrived at Espelette, hometown of Armand David, the scientific discoverer of the giant panda, on July 10, 2014, after 106 days’ journey. He cycled about 15,000km through such countries as Kazakhstan, Russia, Belarus, Poland, and Germany. He has been called “China’s Giant Panda Cyclist” for his promotion of panda culture all the way to France.

左 2018 年 1 月，全世界第一部大熊猫志书——《四川省志 • 大熊猫志》，方志出版社出版
右 2019 年 6 月，全世界第一部《大熊猫图志》，方志出版社出版

(Left) *Sichuan Province Annals: The Giant Panda*, the world's first giant panda annals was published by the Publishing House of Local Records in January 2018.

(Right) *Giant Panda Pictorial Annals*, the world's first giant panda pictorial annals was published by the Publishing House of Local Records in June 2019.

《四川省志 • 大熊猫志》《大熊猫图志》

2018 年 1 月，四川省地方志编纂委员会编纂、方志出版社出版的《四川省志·大熊猫志》面世，它是全球首部全方位记录大熊猫的官修志书。该书编纂审核历时 12 年，向世人揭示了大熊猫的演化历史和科学发现后百余年的故事。2019 年 6 月，四川省地方志工作办公室、四川省林业和草原局联合编著出版《大熊猫图志》。

Sichuan Province Annals: Giant Panda, and *Giant Panda Pictorial Annals*

In January 2018, *Sichuan Province Annals: The Giant Panda*, compiled by Sichuan Local Records Editorial Committee, was published by the Publishing House of Local Records. Having gone through 12 years' of preparation, this document is the world's first official giant panda annals that gives a comprehensive record of this species by telling its history of evolution, and stories over the past 150 years since its scientific discovery. Then, in June 2019, *Giant Panda Pictorial Annals* came out, jointly compiled by The Office of Sichuan Province Local Records and Sichuan Province Forestry and Grassland Administration.

与大熊猫相关的部分研究和文化作品（2019 年） 考拉看看 供图
Some monographs and cultural works concerning the giant panda (2019). Photo courtesy of Koalacan.

文化和研究作品

全世界关于大熊猫的文学、摄影和艺术作品数不胜数。如第 29 届奥林匹克运动会组织委员会、四川省人民政府赠书《最爱献给全世界》，胡锦矗和夏勒合著的《卧龙的大熊猫》，魏辅文的《野生大熊猫科学探秘》，黄万波与魏光飚的《大熊猫的起源》，谭楷的《熊猫的故事》，周孟棋的《中国大熊猫》，童昌信的大熊猫百米水墨长卷《生生不息》等。这些大熊猫研究专家、作家和艺术家为大熊猫文化传播做出了十分重大的贡献。

Giant Panda Cultural Works and Treatises

There are numerous literary works, albums, artistic works and treatises concerning the giant panda across the world. They include *The Giant Panda: A Favorite with Mankind* (government gift book) by Beijing Organizing Committee for the 29th Olympic Games and the Government of Sichuan Province; *The Giant Panda of Wolong* by Hu Jinchu, George B. Schaller, et al., *Scientific Exploration of the Wild Giant Panda* by Wei Fuwen; *The Origin of Giant Panda* by Huang Wanbo and Wei Guangbiao; *A Story of the Giant Panda* by Tan Kai; *China's Giant Panda* by Zhou Mengqi; and *Circle of Life* by Tong Changxin, a 100-meter-long Chinese scroll painting featuring the giant panda. These writers and artists have all contributed significantly to the publicity of giant panda culture.

亚成体期大熊猫正在攀爬、玩耍 引自四川省档案局、四川省档案学校编《档案见证·大熊猫》，四川人民出版社，2018 年

A sub-adult giant panda is having fun by climbing. Cited from *Archives Witness: The Giant Panda* compiled by Sichuan Province Archives Bureau and Sichuan Archives College, and published by Sichuan People's Publishing House, 2018

姐妹城市

2000年11月13日，阿尔芒戴维逝世100周年之际，其家乡法国埃斯佩莱特市的45位市民在市长安德烈戴海杜的率领下，访问四川省宝兴县邓池沟天主教堂，受到宝兴县人民政府热情接待。2002年，四川省雅安市人民政府副市长孙前率团回访，并与埃斯佩莱特市政府共同签署宝兴—埃斯佩莱特结为姐妹城市的友好协议。中法两国的两座城市，因为大熊猫结下深厚友谊。

Sister Cities

On November 13, 2000, for the 100th anniversary of David's death, 45 citizens from David's hometown Espelette, led by Mayor Andre Darraidou, visited Dengchigou Catholic Church in Baoxing County, Sichuan, and were warmly received by the local government. In 2002, Sun Qian, deputy mayor of Ya'an City, Sichuan, heading a delegation, paid a return visit and signed a goodwill agreement between Baoxing and Espelette as sister cities, thanks to the giant panda.

大美

美国动物学家乔治·夏勒在《最后的熊猫》[1]一书中这样评价大熊猫："它虽然体格肥胖，行动迟缓，却独具创作的天分、艺术的完美，仿佛专门为了这项崇高的目标而进化成这种模样。圆圆的扁脸，描黑圈的大眼睛，圆滚滚逗人想抱的体型，赋予熊猫一种天真、孩子气的特质，赢得所有人的怜爱，想要抱它、保护它，而且它又很罕见。"这也解释了大熊猫热为何在全球各国一浪高过一浪，经久不衰。

Beauty

George B. Schaller, an American zoologist, made the following comments on the giant panda in his treatise, *The Last Panda*,[1] "Though dumpy and bearlike, it has been patterned with such creative flourish, such artistic perfection, that it almost seems to have evolved for this higher purpose. A round, rather flat face, large black eye patches, and a cuddly and clumsy appearance give the panda an innocent, childlike quality that evokes universal empathy, a desire to hug and protect it. And it is so rare, too." This explains why the giant panda craze keeps sweeping the world.

[1] 乔治·夏勒．最后的熊猫 [M]．张定绮，译．北京：光明日报出版社，1998. 11.

[1] George B. Schaller, *The Last Panda* [M]. Chicago and London: The University of Chicago Press, 1993.

好吗？（2016 年） 周孟棋 摄
How are you? (2016) Photo by Zhou Mengqi

附录

参考文献

[1] 四川省地方志编纂委员会．四川省志·大熊猫志 [M]．北京：方志出版社，2018. 1.

[2] 四川省地方志编纂委员会．四川省志·大事记述 [M]．成都：四川科学技术出版社，1999. 12.

[3] 四川省地方志编纂委员会．四川省志·旅游志 (1986—2005) [M]．北京：方志出版社，2014. 9.

[4] 四川省地方志编纂委员会．四川省志·林业志 (1986—2005) [M]．北京：方志出版社，2014. 12.

[5] 四川省地方志编纂委员会．四川省志·体育志 (1986—2005) [M]．北京：方志出版社，2018. 6.

[6] 陕西省地方志编纂委员会．陕西省志·林业志 [M]．北京：中国林业出版社，1996. 8.

[7] 甘肃省地方史志编纂委员会．甘肃省志·林业志 [M]．兰州：甘肃人民出版社，1999. 4.

[8] 甘肃省地方史志编纂委员会．甘肃省志·旅游志 [M]．兰州：甘肃文化出版社，2007. 11.

[9] 四川经济年鉴编辑委员会．四川年鉴 (1986[M]．成都：四川科学技术出版社，1987. 8.

[10] 四川年鉴社．四川年鉴 (2013) [M]．成都：四川年鉴社，2013. 12.

[11] 四川年鉴社．四川年鉴 (2016) [M]．成都：四川年鉴社，2016. 12.

[12] 四川年鉴社．四川年鉴 (2017) [M]．成都：四川年鉴社，2017. 12.

[13] 陕西年鉴编纂委员会．陕西年鉴 (2016) [M]．西安：陕西年鉴社，2016. 10.

[14] 陕西年鉴编纂委员会．陕西年鉴 (2017) [M]．西安：陕西年鉴社，2017. 10.

[15] 陕西年鉴编纂委员会．陕西年鉴 (2018) [M]．西安：陕西年鉴编辑部，2018. 11.

[16] 甘肃省地方史志办公室．甘肃年鉴 (2016) [M]．北京：中国文史出版社，2016. 11.

[17] 甘肃省地方史志办公室．甘肃年鉴 (2017) [M]．北京：中国文史出版社，2017. 10.

[18] 甘肃省地方史志办公室．甘肃年鉴 (2018) [M]．兰州：甘肃民族出版社，2018. 10.

[19] 四川省唐家河国家级自然保护区管理处，四川省广元唐家河风景区管理局．四川唐家河国家级自然保护区志（公元前 201—公元 2010）[M]．成都：四川科技出版社，2012. 12.

[20] 汶川特大地震四川抗震救灾志编纂委员会．汶川特大地震四川抗震救灾志 [M]．成都：四川人民出版社，2018. 3.

[21] 成都市地方志编纂委员会．成都市志·园林志 [M]．成都：四川人民出版社，1998. 10.

[22] 凉山彝族自治州地方志编纂委员会．凉山彝族自治州志(1991—2006)[M]．北京:方志出版社，2011. 8.

[23] 甘孜藏族自治州地方志编纂委员会．甘孜州志（1991—2005）[M]．成都：四川人民出版社，2010. 9.

[24] 广元市地方志编纂委员会．广元市志（1985—2004）[M]．北京：方志出版社，2016. 7.

[25] 西昌市志编纂委员会．西昌市志（1999—2005）[M]．北京：方志出版社，2011. 8.

[26] 石棉县地方志编纂委员会．石棉县志（1986—2000）[M]．北京：中国文史出版社，2010. 11.

[27] 九寨沟县地方志编纂委员会．九寨沟县志（1986—2005）[M]．北京：方志出版社，2011. 5.

[28] 宝兴县志编纂委员会．宝兴县志 [M]．北京：方志出版社，2000. 11.

[29] 佛坪县地方志编纂委员会．佛坪县志 [M]．西安：三秦出版社，1993. 12.

[30] 文县志编纂委员会．文县志 [M]．兰州：甘肃人民出版社，1997. 12.

[31] 舟曲县地方史志编纂委员会．舟曲县志（1991—2006）[M]．北京：方志出版社，2010. 12.

[32] 迭部县志编纂委员会．迭部县志 [M]．兰州：兰州大学出版社，1998. 6.

[33] 李晟之、蒋仕伟．王朗志：四川王朗国家级自然保护区志（1965—2015）[M]．成都：四川科学技术出版社，2018. 8.

[34] 四川省档案局、四川省档案学校．档案见证•大熊猫 [M]．成都：四川人民出版社，2018. 3.

[35] 孙前．大熊猫文化笔记 [M]．北京：五洲传播出版社，2009. 11.

[36] 黄万波、魏光飚．大熊猫的起源 [M]．北京：科学出版社，2010. 5.

[37] 万本根、蒲涛、石恒方．最爱献给全世界——大熊猫摄影作品集 [M]．成都:四川民族出版社，2008. 7.

[38] 魏辅文．野生大熊猫科学探秘 [M]．北京：科学出版社，2018. 2.

[39] 谭楷．熊猫的故事 [M]．北京：五洲传播出版社，2015. 1.

[40] 四川省林业厅．四川的大熊猫：四川省第四次大熊猫调查报告 [M]．成都：四川科学技术出版社，2015. 12.

[41] 胡锦矗．追踪大熊猫的岁月 [M]．郑州：海燕出版社，2005. 12.

[42] 陈能文．大熊猫世界 [M]．成都：四川民族出版社，1993. 7.

[43] 冯文和、李光汉．拯救大熊猫 [M]．成都：四川科学技术出版社，2000. 12.

[44] 乔治•夏勒．最后的熊猫 [M]．张定绮，译．北京：光明日报出版社，1998. 11.

[45] 阿尔芒•戴维．戴维日记 [M]．海伦•佛克斯，译．波士顿：哈佛大学出版社，1949.

Appendices

Bibliography

[1] Sichuan Local Records Editorial Committee. *Sichuan Province Annals: Giant Panda* [M]. Beijing: Publishing House of Local Records, 2018.

[2]Sichuan Local Records Editorial Committee. *Sichuan Province Annals: Tourism (1986-2005)* [M]. Beijing: Publishing House of Local Records, 2014.

[3]Sichuan Local Records Editorial Committee. *Sichuan Province Annals: Chronicle of Events* [M]. Chengdu: Sichuan Science and Technology Press, 1999.

[4]Sichuan Local Records Editorial Committee. *Sichuan Province Annals: Forestry (1986-2005)* [M]. Beijing: Publishing House of Local Records, 2014.

[5] Sichuan Local Records Editorial Committee. *Sichuan Province Annals: Sports (1986-2005)* [M]. Beijing: Publishing House of Local Records, 2018.

[6]Shaanxi Local Records Editorial Committee. *Shaanxi Province Annals: Forestry* [M]. Beijing: China Forestry Publishing House, 1996.

[7] Gansu Local Records Editorial Committee. *Gansu Province Annals: Forestry* [M]. Lanzhou: Gansu People's Publishing House, 1999.

[8] Gansu Local Records Editorial Committee. *Gansu Province Annals: Tourism* [M]. Lanzhou: Gansu Culture Press, 2007.

[9] Editorial Board of Sichuan Economic Yearbook. *Sichuan Yearbook (1986)* [M]. Chengdu: Sichuan Science and Technology Press, 1987.

[10] Sichuan Yearbook Press. *Sichuan Yearbook (2013)* [M]. Chengdu: Sichuan Yearbook Press, 2013.

[11] Sichuan Yearbook Press. *Sichuan Yearbook (2016)* [M]. Chengdu: Sichuan Yearbook Press, 2016.

[12] Sichuan Yearbook Press. *Sichuan Yearbook (2017)* [M]. Chengdu: Sichuan Yearbook Press, 2017.

[13] Shaanxi Yearbook Editorial Committee. *Shaanxi Yearbook (2016)* [M]. Xi'an: Shaanxi Yearbook Press, 2016.

[14] Shaanxi Yearbook Editorial Committee. *Shaanxi Yearbook (2017)* [M]. Xi'an: Shaanxi Yearbook Press, 2017.

[15] Shaanxi Yearbook Editorial Committee. *Shaanxi Yearbook (2018)* [M]. Xi'an: Shaanxi Yearbook Editorial Office,

2018.

[16] Gansu Local Records Office. *Gansu Yearbook (2016)* [M]. Beijing: China Culture and History Press, 2016.

[17] Gansu Local Records Office. *Gansu Yearbook (2017)* [M]. Beijing: China Culture and History Press, 2017.

[18] Gansu Local Records Office. *Gansu Yearbook (2018)* [M]. Lanzhou: Gansu Ethnic Press, 2018.

[19] Sichuan Tangjiahe National Nature Reserve Management Office, Sichuan Guangyuan Tangjiahe Scenic Spot Administration. *Sichuan Tangjiahe National Nature Reserve Records (201 B.C.–2010 A.D.)* [M]. Chengdu: Sichuan Science and Technology Press, 2012.

[20]Editorial Board of Records of Sichuan Earthquake Relief after Wenchuan Massive Earthquake. *Records of Sichuan Earthquake Relief after Wenchuan Massive Earthquake* [M]. Chengdu: Sichuan People's Publishing House, 2018.

[21] Chengdu Local Records Editorial Committee. *Chengdu Local Records: Horticulture* [M]. Chengdu: Sichuan People's Publishing House, 1998.

[22] Editorial Board of Liangshan Yi Autonomous Prefecture Local Records. *Liangshan Yi Autonomous Prefecture Local Records (1991-2006)* [M]. Beijing: Publishing House of Local Records, 2011.

[23] Editorial Board of Ganzi Tibetan Autonomous Prefecture Local Records. *Ganzi Prefecture Local Records (1991-2005)* [M]. Chengdu: Sichuan People's Publishing House, 2010.

[24] Guangyuan Local Records Editorial Committee. *Guangyuan Local Records (1985-2004)* [M]. Beijing: Publishing House of Local Records, 2016.

[25] Xichang Local Records Editorial Committee. *Xichang City Local Records (1999-2005)* [M]. Beijing: Publishing House of Local Records, 2011.

[26] Shimian County Local Records Editorial Committee. *Shimian County Local Records (1986-2000)* [M]. Beijing: China Culture and History Press, 2010.

[27] Jiuzhaigou County Local Records Editorial Committee. *Jiuzhaigou County Local Records (1986-2005)* [M]. Beijing: Publishing House of Local Records, 2011.

[28] Baoxing County Local Records Editorial Committee. *Baoxing County Local Records* [M]. Beijing: Publishing House of Local Records, 2000.

[29] Foping County Local Records Editorial Committee. *Foping County Local Records* [M]. Xi'an: Sanqin Publishing House, 1993.

[30] Wenxian County Local Records Editorial Committee. *Wenxian County Local Records* [M]. Lanzhou: Gansu People's Publishing House, 1997.

[31] Zhouqu County Local Records Editorial Committee. *Zhouqu County Local Records (1991-2006)* [M]. Beijing: Publishing House of Local Records, 2010.

[32] Têwo County Local Records Editorial Committee. *Têwo County Local Records* [M]. Lanzhou: Lanzhou University Press, 1998.

[33] Li Shengzhi, Jiang Shiwei. *Local Records of Sichuan Wanglang National Nature Reserve (1965-2015)* [M]. Chengdu: Sichuan Science and Technology Press, 2018.

[34] Sichuan Province Archives Bureau, *Sichuan Province Archives College. Archives Witness: The Giant Panda* [M]. Chengdu: Sichuan People's Publishing House, 2018.

[35] Sun Qian. *Notes of the Giant Panda Culture* [M]. Beijing: China Intercontinental Press, 2009.

[36] Huang Wanbo, Wei Guangbiao. *The Origin of Giant Panda* [M]. Beijing: Science Press, 2010.

[37] Wan Bengen, Pu Tao, Shi Hengfang. *Giant Panda A Favorite with Mankind* [M]. Chengdu: Sichuan Ethnic Press, 2008.

[38]Wei Fuwen. *Scientific Exploration of Wild Giant Pandas* [M]. Beijing: Science Press, 2018.

[39] Tan Kai. *Story of the Giant Panda* [M]. Beijing: China Intercontinental Press, 2015.

[40] Forestry Department of Sichuan Province. *The Pandas of Sichuan The 4th Survey Report on Giant Pandas in Sichuan Province* [M]. Chengdu: Sichuan Science and Technology Press, 2015.

[41] Hu Jinchu. *The Years of Tracking Giant Pandas* [M]. Zhengzhou: Petrel Publishing House, 2005.

[42] Chen Nengwen. *The World of Giant Pandas* [M]. Chengdu: Sichuan Ethnic Press, 1993.

[43] Feng Wenhe, Li Guanghan. *The Saving of Giant Panda* [M]. Chengdu: Sichuan Science and Technology Press, 2000.

[44] George B. Schaller, *The Last Panda* [M]. Chicago and London: The University of Chicago Press, 1993.

[45] Armand David, *Abbe David's Diary* [M]. Trans. Helen M. Fox. Boston: Harvard University Press, 1949.

附录

重要文件

表附

时间	单位	文件名
1950 年 5 月	中央人民政府政务院	《古迹、珍贵文物、图书及稀有生物保护办法并古文化遗址及古墓葬之调查发掘暂行办法》
1956 年 10 月	林业部	《关于天然森林禁伐区（自然保护区）划定方案》
1959 年 2 月	林业部	《关于积极开展狩猎事业的指示》
1962 年 9 月	国务院	《关于积极保护和合理利用野生动物资源的指示》
1975 年 3 月	国务院批转农林部 四川省革命委员会	《关于四川省珍贵稀有动物保护管理情况的调查报告》
1976 年 3 月	农林部	《关于加强大熊猫保护工作的紧急通知》
1978 年 12 月	国务院转发国家林业总局	《关于加强大熊猫保护、驯养工作的报告》
1979 年 3 月	四川省革命委员会	《关于加强自然保护区建设的通知》
1981 年 1 月	四川省人民政府	《关于加强野生动物资源保护和狩猎管理的布告》
1983 年 4 月	国务院	《关于严格保护珍贵、稀有野生动物的通令》
1983 年 8 月	国务院转发林业部	《关于抢救大熊猫的紧急报告的通知》
1983 年 9 月	四川省人民政府办公厅批转省林业厅	《关于落实抢救大熊猫措施的紧急报告》
1984 年 4 月	四川省人民政府	《关于进一步做好抢救、保护大熊猫工作的通知》
1984 年 6 月	四川省人民代表大会常务委员会	《关于抢救和保护大熊猫的决议》
1985 年 7 月	林业部	《森林和野生动物类型自然保护区管理办法》
1987 年 2 月	四川省人民政府	《关于切实做好保护、抢救大熊猫工作的通知》

续表附

时间	单位	文件名
1987 年 7 月	最高人民法院	《关于依法严惩猎杀大熊猫，倒卖、走私大熊猫皮的犯罪活动的通知》
1987 年 8 月	国务院	《关于坚决制止乱捕滥猎和倒卖、走私珍稀野生动物的紧急通知》
1987 年 9 月	四川省人民政府	《贯彻国务院〈关于坚决制止乱捕滥猎和倒卖、走私珍稀野生动物的紧急通知〉的通知》
1988 年 11 月	全国人民代表大会常务委员会	《关于惩治捕杀国家重点保护的珍贵、濒危野生动物犯罪的补充规定》
1988 年 12 月	国务院	《关于〈国家重点保护野生动物名录〉的批复》
1989 年 10 月	林业部与世界自然基金会	《中国大熊猫及其栖息地保护管理计划》
1990 年 1 月	四川省第七届人大常务委员会	《四川省〈中华人民共和国野生动物保护法〉实施办法》
1990 年 3 月	四川省人民政府	《四川省重点保护野生动物名录》
1992 年 1 月	四川省人民政府办公厅转发省保护抢救大熊猫领导小组	《关于进一步做好保护抢救大熊猫工作的意见》
1992 年 2 月	国务院	《中华人民共和国陆生野生动物保护实施条例》
1994 年 9 月	国务院	《中华人民共和国自然保护区条例》
1999 年 10 月	四川省人民代表大会常务委员会	《四川省自然保护区管理条例》
2001 年 2 月	国家林业局	《野外大熊猫救护工作规定》
2002 年 7 月	四川省人民政府批复省林业厅	《四川省野生动植物保护及自然保护区建设总体规划》
2005 年 6 月	四川省野生动物资源调查保护管理站	《全国大熊猫野化放归总体规划（2005—2014 年）》
2017 年 1 月	中共中央办公厅、国务院办公厅	《大熊猫国家公园体制试点方案》
2017 年 8 月	四川省大熊猫国家公园体制试点工作推进领导小组	《大熊猫国家公园体制试点实施方案（2017—2020 年）》
2018 年 10 月	四川省林业厅	《四川省大熊猫野化放归技术指南（2018—2027）》

Appendices

Key Documents

In Table Attached

Time	Department	Documents
May 1950	The Administration Council of the Central People's Government	*Interim Measures for the Protection of Historical Sites, Precious Cultural Relics, Books, and Rare Animals and Plants, and for the Investigation and Excavation of Ancient Cultural Sites and Ancient Mausoleums*
October 1956	Ministry of Forestry	*Plan on the Delimitation of Protected Forest Areas (Nature Reserves)*
February 1959	Ministry of Forestry	*Directives on Hunting*
September 1962	State Council	*Directives on Conservation and Rational Utilization of Wildlife Resources*
March 1975	Approved and forwarded to the Ministry of Agriculture and Forestry by the State Council, Sichuan Province Revolutionary Committee	*Investigation Report on the Conservation and Management of Rare and Precious Animals in Sichuan Province*
March 1976	Ministry of Agriculture and Forestry	*Emergency Notice on Strengthening the Protection of Giant Pandas*
December 1978	Forwarded to the State Forestry Administration by the State Council	*Report on Strengthening the Protection and Domestication of Giant Pandas*
March 1979	Sichuan Province Revolutionary Committee	*Notice on Strengthening the Construction of Nature Reserves*
January 1981	People's Government of Sichuan Province	*Notice on Strengthening Protection of Wildlife Resources and Management of Hunting*
April 1983	State Council	*Decree on Strict Protection of Rare and Precious Wild Animals*
August 1983	Forwarded to the Ministry of Forestry by The State Council	*Notice on Emergency Report on the Rescue of Giant Pandas*

Continued

Time	Department	Documents
September 1983	Approved and forwarded to the Forestry Department of Sichuan Province by The General Office of the People's Government of Sichuan Province	*Emergency Report on the Implementation of Rescue Measures for Giant Pandas*
April 1984	People's Government of Sichuan Province	*Notice on Further Rescue and Protection of Giant Pandas*
June 1984	Standing Committee of Sichuan Provincial People's Congress	*Resolution on the Rescue and Protection of Giant Pandas*
July 1985	Ministry of Forestry	*Measures for the Administration of Forest and Wildlife Type Nature Reserves*
February 1987	People's Government of Sichuan Province	*Notice on the Effective Protection and Rescue of Giant Pandas*
July 1987	Supreme People's Court	*Notice on Punishment by law of the Criminal Activities of Hunting and Killing Giant Pandas, Buying, Reselling and Smuggling Giant Panda Pelts*
August 1987	State Council	*Emergency Notice on the Prohibition of Hunting, Buying, Reselling, and Smuggling Rare Wild Animals*
September 1987	People's Government of Sichuan Province	*Notice on the implementation of The State Council's Emergency Notice on the Prohibition of Hunting, Buying, Reselling, and Smuggling Rare Wild Animals*
November 1988	People's Government of Sichuan Province	*Supplementary Provisions on the Punishment of the Criminal Activities of Hunting and Killing Rare and Endangered Wildlife Under Special State Protection*
December 1988	State Council	*Reply to The List of Wildlife Under Special State Protection*
October 1989	Ministry of Forestry and WWF	*China Conservation and Management Plan for Giant Pandas and Habitats*
January 1990	Standing Committee of the Seventh People's Congress of Sichuan Province	*Sichuan Provincial Measures for the Implementation of Law of the People's Republic of China on the Protection of Wildlife*

Continued

Time	Department	Documents
March 1990	People's Government of Sichuan Province	*List of Wildlife under Special Protection in Sichuan Province*
January 1992	Forwarded to the Provincial Leading Group on Giant Panda Conservation and Rescue by The General Office of The People's Government of Sichuan Province	*Opinions on Further Protection and Rescue of Giant Pandas*
February 1992	State Council	*Regulations of the People's Republic of China on the Implementation of Land Wildlife Protection*
September 1994	State Council	*Regulations of the People's Republic of China on Nature Reserves*
October 1999	Standing Committee of Sichuan Provincial People's Congress	*Sichuan Provincial Regulations on the Administration of Nature Reserves*
February 2001	State Forestry Administration	*Provisions on Rescue Work for Wild Giant Pandas*
July 2002	Approved and replied to Sichuan Provincial Forestry Department by The People's Government of Sichuan Province	*Sichuan Province General Plan for Wildlife Protection and Nature Reserve Construction*
June 2005	Sichuan Province Wild Animal Resources Investigation and Protection Management Station	*The Master Plan for Giant Panda Reintroduction (2005-2014)*
January 2017	The General Office of the Central Committee of the Communist Party of China and the General Office of the State Council of the People's Republic of China	*The Pilot Program of Giant Panda National Park System*
August 2017	The promotion leadership group of the Pilot Program of Sichuan Giant Panda National Park System	*The Pilot Program of Giant Panda National Park System (2017-2020)*
October 2018	Sichuan Province Forestry Administration	*A Technical Guideline for Giant Panda Reintroduction in Sichuan Province (2018-2027)*

后记

记录历史　守望未来

这本图志是记录历史，也是守望未来。

中国人对记录保有持续的兴趣，从甲骨文到造纸术，从今天这本创新的图志到它记录的主角大熊猫，世界变化，万象更新。

2019 年是大熊猫科学发现 150 周年，此前作为明星的大熊猫从来不乏记录者，我们为何还要继续？因为文明的种子需有人持续播种，真正的热爱要付诸行动。

大熊猫的科学价值经过 150 年的发酵，已经足够醇香，而它的文化和商业价值之门才刚刚打开。开大众之智，助众生见识，是为大道。

这是一次全新的尝试，旨在以通俗图志传播大熊猫价值。这本图志从创意到实施，从选图到设计，从制作到印刷，我们得到诸多帮助，借此一并致谢！

我们要特别感谢西华师范大学、中国科学院动物研究所、四川大学、中国大熊猫保护研究中心、成都大熊猫繁育研究基地、四川省档案馆、重庆市地质调查院、四川省地方志系统、四川省林业和草原系统、四川卧龙国家级自然保护区管理局、考拉看看团队、书服家团队、四川省文艺期刊联合会《现代艺术》杂志社、中国驻西班牙大使馆、西班牙马德里动物园等单位（或系统）；我们要特别感谢西华师范大学珍稀动植物研究所胡锦矗教授、中国科学院院士魏辅文研究员、中国大熊猫保护研究中心常务副主任张和民教授、成都大熊猫繁育研究基地主任张志和教授、西华师范大学珍稀动植物研究所张泽钧研究员、四川大学生命科学学院副院长冉江洪教授、重庆市地质调查院副院长魏光飚研究员、四川省人民政府原副秘书长薛康、原四川省旅游局巡视员孙前、大熊猫文化学者谭楷、资深摄影家周孟棋、高华康等。受篇幅限制，这里未一一列明机构和个人的名字。我们要向这些对大熊猫关爱至极的专家、学者表达敬佩之情，他们为编著出版《大熊猫图志》提供了巨大的帮助和支持！

此外，在面向全社会开展的图片征集活动中，许多热心市民积极参与，他们将相册里珍藏多年的宝贵照片、画框里价值不菲的作品无私地贡献出来，为我们的编著工作倾力相助。

正是在大家的鼓励和帮助下，《大熊猫图志》才能顺利出版。由于编者学识有限，对一些问题的认识尚有不足，敬请批评指正。

再次感谢读者阅读本书！

四川省地方志工作办公室
四川省林业和草原局
2019 年 6 月 30 日

Epilogue

From Past to Future

This document—*Giant Panda Pictorial Annals*—records the past and looks forward to the future.

The Chinese people have shown a great interest in record keeping, from oracle bone inscriptions to the art of papermaking, and to this special pictorial annals featuring the giant panda.

The year of 2019 marks the 150th anniversary of the scientific discovery of the giant panda. Why is it a necessity for us to continue recording the giant panda as there are already many records about this species? Because the torch of civilization needs to be relayed, and actions need to be taken to show our compassion and enthusiasm.

We have been studying the giant panda for 150 years, a period long enough for us to accumulate a scientific wealth of this species. However, its cultural value has yet to be explored. Therefore, we take it as our duty to inform our readers of everything related to the giant panda, both scientific and cultural.

This book is a new attempt to popularize the value of the giant panda. We'd like to take this chance to acknowledge those who have been of great help to us in providing and actualizing original ideas, in selecting photos and designing the layout, and in getting the book published.

Special thanks should go to China West Normal University; Institute of Zoology of the Chinese Academy of Sciences (CAS); Sichuan University; China Conservation and Research Center for Giant Panda; Chengdu Research Base of Giant Panda Breeding; Sichuan Province Archives; Chongqing Institute of Geological Survey; all local records offices in Sichuan; all branches of Sichuan Province Forestry and Grassland Administration; and Sichuan Wolong National Nature Reserve Administration; Koalacan Team; Forbooks Team; *Modern Art* magazine of Sichuan Federation of Literary Journals; China Embassy in Spain; Madrid Zoo Aquarium, among others. We'd also like to extend our great gratitude to Professor Hu Jinchu from Institute of Rare Animals and Plants, China West Normal University; Wei Fuwen, academician with Chinese Academy of Sciences; Professor Zhang Hemin, executive deputy director of China Conservation and Research Center for Giant Panda; Professor Zhang Zhihe, director of Chengdu Research Base of Giant Panda Breeding;Zhang Zejun, researcher from Institute of Rare Animals and Plants of China West Normal University; Professor Ran Jianghong, vice dean of College of Life Sciences from Sichuan University; Wei Guangbiao, researcher and vice president of Chongqing Institute of Geological Survey; Xue Kang, former deputy secretary-general of People's Government of Sichuan Province; Sun Qian, former consultant of Sichuan Tourism Administration; Tan Kai, scholar of giant panda culture; Zhou Mengqi and Gao Huakang, senior photographers. Though we've tried to list all the institutions and individuals who have provided us with help and support for this document, there are still some experts and scholars who have shown great care and concern for the giant panda we have failed to include but so appreciated.

In addition, when we were collecting giant panda pictures from the public, many citizens gladly offered us their time-honored precious photos.

This book owes its publication to everyone's effort and support. As there might be limitations and drawbacks, we shall truly appreciate any advice and suggestions.

Thank you for your time!

The Office of Sichuan Province Local Records

Sichuan Province Forestry and Grassland Administration

June 30, 2019

水彩画《杰瑞和小馨》，27cm×19.5cm（2017 年） 曹元元 作品

Watercolor painting: *Jerry and Xiao Xin*, 27cm×19.5cm (2017). By Cao Yuanyuan

剪纸《合家欢》，50cm×150cm（2014 年） 袁成祥 作品

Paper-cutting:*Happy Family*, 50cm×150cm (2014). By Yuan Chengxiang

大熊猫图志

GIANT PANDA PICTORIAL ANNALS

八百万年来，大熊猫适应地质变迁和气候变化，繁衍至今，成为动物界极为罕见的“活化石”，更成为全世界的超级明星。2019年是大熊猫科学发现150周年，在四川省人民政府新闻办公室指导下，四川省地方志工作办公室、四川省林业和草原局联合编著出版全球首部大熊猫图志，旨在以通俗简洁的语言、400余幅精美图照，全方位展示大熊猫的起源、生态、研究、保护和文化。

翻开此书，我们不仅能触摸到大熊猫的生活百态，感受到人与大熊猫的友好亲近，还能体悟到科学家们的不懈探索与孜孜追求，更能领略到自然的神奇与造物的伟大。

Giant Pandas, since eight million years ago, have outlived all geological and climate changes with their reproduction as the “living fossil” of the animal kingdom and celebrity status around the world. As the year 2019 marks the 150th anniversary of its scientific discovery, The Office of Sichuan Province Local Records and Sichuan Province Forestry and Grassland Administration have jointly compiled and published the world’s first Giant Panda Pictorial Annals under the guidance of Information Office of Sichuan Provincial People’s Government. *GIANT PANDA PICTORIAL ANNALS* aims at illustrating in simple and concise language giant pandas’ origin, ecology, research, protection and culture, through over 400 pictures.

Opening this book, our readers can feel not only the facets of these creatures through friendly encounters with them, but also the wonder of nature and greatness of its Creator, thanks to all the scientists’ painstaking efforts.

指导单位 四川省人民政府新闻办公室

编 著 四川省地方志工作办公室
四川省林业和草原局

Guiding Board Information Office of Sichuan Provincial People’s Government

Edited by The Office of Sichuan Province Local Records
Sichuan Province Forestry and Grassland Administration

读者服务 / Reader Service

4000213677 (028)84525271